FATAL ATTRACTION

Magnetic Mysteries of the Enlightenment

FATAL ATTRACTION

Magnetic Mysteries of the Enlightenment

Patricia Fara

REVOLUTIONS IN SCIENCE
Published by Icon Books UK

Published in the UK in 2005
by Icon Books Ltd., The Old Dairy,
Brook Road, Thriplow,
Cambridge SG8 7RG
email: info@iconbooks.co.uk
www.iconbooks.co.uk

Sold in the UK, Europe, South Africa
and Asia by Faber and Faber Ltd.,
3 Queen Square, London WC1N 3AU
or their agents

Distributed in the UK, Europe, South Africa
and Asia by TBS Ltd., Frating Distribution Centre,
Colchester Road, Frating Green, Colchester CO7 7DW

Published in Australia in 2005
by Allen & Unwin Pty. Ltd.,
PO Box 8500, 83 Alexander Street,
Crows Nest, NSW 2065

Distributed in Canada by
Penguin Books Canada,
10 Alcorn Avenue, Suite 300,
Toronto, Ontario M4V 3B2

ISBN 1 84046 632 4

Series editor: Jon Agar

Originating editor: Simon Flynn

Typesetting by Hands Fotoset

Printed and bound in the UK by Mackays of Chatham plc

Dedication
For Nick and Nathalie

ACKNOWLEDGEMENTS

Because readers allegedly dislike footnotes, I have mostly restricted references to direct quotations. However, I have of course drawn on a great number of scholarly books, as well as being helped by many colleagues while I was carrying out my research. So I should like to emphasise how grateful I am for all the information about magnetism and its history that I have read and been given since 1989, when I first started work on this topic. I also want to thank all those people who have commented on my writing and lecturing: it is a great privilege to belong to a scholarly community in which comments by students can be as valuable as those by professors. Just as importantly, I should like to thank Simon Flynn and Andrew Furlow at Icon for all their encouragement, energy and enthusiasm.

CONTENTS

LIST OF ILLUSTRATIONS

MYSTERIOUS MAGNETS

... by a power
Like the polar Paradise
Magnet-like of lovers' eyes;
I, a most enamoured maiden
Whose weak brain is overladen
With the pleasure of her love,
Maniac-like around thee move ...

Percy Bysshe Shelley, *Prometheus Unbound*, 1820

And so to bed ... In Samuel Pepys's diary, his famous catch-phrase marked the end of a busy day. But sometimes Pepys found it hard to sleep, and science was one of the things that kept him awake. Once he sat up into the small hours studying a book about the new microscope that his friend Robert Hooke had invented. Like readers all over the country, Pepys marvelled at Hooke's detailed pictures of the fleas and lice that pestered even the wealthiest gentlemen of the 17th century. Ironically, the following year plague swept through London's crowded homes, but neither Hooke nor Pepys suspected that these tiny insects were the culprits.

Amid the sex and the scandals that Pepys secretly noted down in shorthand, he found time to ponder Hooke's attempts to measure the speed of a fly's flapping wings and to marvel at a successful blood transfusion between dogs: what would happen, Pepys wondered subversively, if 'the blood of a Quaker be let into an Archbishop'? He also met a human guinea-pig who reported (in Latin) feeling much better after receiving some sheep's blood, although – Pepys observed – he had been bribed with a pound and was in any case 'cracked a little in his head'.

Hooke's search for knowledge had many critics. How useful can it be, demanded the Duchess of Newcastle, to produce magnified pictures of bees? Surely what we need is more honey? But like the Duchess, Pepys, Hooke and their colleagues believed that research should produce practical results. Pepys had faith that scientific investigations would bring if not fame and fortune, then at least a better future. He was fascinated by some 'magnetique experiments' that he watched with a group of friends; as a naval man, he knew that better magnetic compasses could help to save lives, ships and goods.[1]

Twenty years later, long after he had stopped consigning his intimate thoughts to his diary, Pepys acquired the highest scientific position in Europe: he became President of London's Royal Society. This was an honorary, unpaid position and Pepys probably already had his eye on the top job at the Admiralty, which commanded a four-figure salary. He had just returned from a disappointing trip to Tangier, during which he kept another shorthand journal, unfortunately less riveting than his earlier one. Instead of revelling in the exotic African sights, Pepys preferred to moan: he caught a cold, his foot hurt (washing his legs in brandy seems to have helped), and he was disgusted by everybody else's sexual activities (or was he, perhaps, jealous?). He also complained about the ignorance of the young boys recruited to work on the royal ships, and the lack of discipline among their officers. Within months of his return, he was appointed head of the Admiralty.

Pepys was intrigued by science, but his power lay at the

Admiralty. Sailing was far more important than science, and being high up in the Navy was far more prestigious than supervising weekly performances of experiments. England's wealth depended on bringing back raw materials and sending out manufactured goods; surrounded by water, the nation relied on its ships for protection. In contrast, the Royal Society was often satirised as a gentlemanly club where incompetent aristocrats deluded themselves with crackpot schemes. The poet Samuel Butler scathingly described a group of bumbling virtuosos who exclaimed with delight at the elephant their telescope had revealed on the moon – until a level-headed sceptic pointed out that a mouse had got stuck in the tube.

Pepys was President of the Royal Society for two years, which is why his name appears on the title page of science's most famous book, Isaac Newton's *Principia Mathematica*. Rich and influential, Pepys could afford to buy the best instruments, and he browsed through the small shops clustered together near the Society in the narrow alleys along the bank of the Thames. One of the merchants he patronised was Thomas Tuttell, a skilled craftsman who became instrument-maker to the king and boasted about his fine brass measuring devices and other high-quality apparatus.

Tuttell excelled at inventing tactics of self-promotion. He ingeniously persuaded customers to buy advertisements of his own products by selling packs of small playing cards demonstrating how useful his mathematical instruments could be. The seven of spades, reproduced here as Figure 1, illustrated magnets, and so brought together two of the major interests in Pepys's life – scientific research at the Royal Society, and navigation.

Figure 1: Playing card. Thomas Tuttell, *Mathematical Cards*, 1701. © British Museum

Charming as this card might be, it seems bizarre to modern eyes. The cherubs may be acceptable as artistic licence, but what has Mahomet got to do with magnetism? Why the large anchor? And who is the man wielding a pickaxe? Just as Pepys's diaries were incomprehensible before they were translated from his idiosyncratic shorthand into plain English, so too these magnetic symbols need to be decoded.

Tuttell's magnetic language is now opaque. Three hundred years ago, there were no red horseshoe-shaped magnets for children to play with (or to symbolise John Smith's beer), no electromagnets and no theories of magnetic fields. Nor were there any computer memories – this might seem obvious, but their absence points to the most fundamental difference between magnetism then and now: magnetism and electricity had not yet been joined together. That happened only in the 19th century.

The very word 'magnet' now conjures up associations totally alien to Pepys and his contemporaries. They hoped that their research would benefit future generations, but they could, of course, never know for sure. We hold an advantage over them: although travelling back in time to explore their vanished magnetic world is hard, it is not impossible. If the past is another country, then this book is a rough guide for those Enlightenment tourists who are drawn by magnetism's mysterious attraction.

* * * * *

Unravelling the magnetic mysteries of the Enlightenment involves forgetting not only about electronic chips, but also about electromagnetic machines and even current electricity, which were both 19th-century innovations. Magnetism and static electricity were regarded as separate forces of nature, and natural philosophers drew up checklists to compare them. Most obviously, magnetism was a silent power that affected iron and steel but could pass painlessly through humans and animals. In contrast, electricity was bright, crackled – and could hurt or even kill. Some brief attempts were made to connect them. Benjamin Franklin, America's great electrical politician, knew that lightning storms at sea could destroy a ship's compasses, and one English researcher electrified his own son to produce colourful streams of light from a magnet in the boy's hand. Even so, the conclusion was unanimous: the differences far outweighed the similarities.

In addition, magnetism and electricity interested separate groups of people. Magnetic attraction had been known since the time of the Greeks, and the 17th-century experts were maritime men – the instrument-makers who built compasses, and the navigators who needed compasses to guide them across the largely uncharted oceans. Magnetism was old and mysterious, whereas electricity was a new, exciting phenomenon that belonged to natural philosophers. Machines to produce static electricity were invented only in the early 18th century, and they rapidly became (literally) dazzling attractions in the repertoire of stage performers and lecturers. Electricity was born as the big buzz science of the Enlightenment, whereas magnetism only slowly grew into being scientific.

In order to empathise with Pepys and his 'magnetique experiments', forget about gravity and electromagnetism: these were introduced later. Visualise instead a universe governed by secret magnetic powers. Magnetic cosmology was taken very seriously in the 17th century. William Gilbert's world-famous book on magnetism had been published in 1600, and he envisioned God's divine magnetic power binding together the entire universe.[2] When Isaac Newton first grappled with comets' orbits, he chose a magnetic explanation. For his *Principia*, the great book on gravity that he published in 1687, Newton relied on the mathematical laws formulated by Johannes Kepler to describe how the planets revolve around the sun – even though Kepler had believed that they were controlled magnetically.

Twenty years before Newton eventually committed his ideas about gravity to print, he was living at his country home in East Anglia, working on the insights that he had (he claimed) gained when sitting beneath an apple tree in his garden. But before he told the scientific world about his conclusions, magnetism reigned. At the same time as Newton was comparing falling apples with the turning moon, in London John Milton was finalising publication of *Paradise Lost*, in which he poetically described the sky's constellations being controlled by the sun's 'magnetic beam'. And from the papal power-base at Rome, the German Jesuit Athanasius Kircher was publishing *The Magnetic Kingdom of Nature*, his own version of an entire magnetic cosmology.

In some ways cosmic magnetic forces were precursors of universal gravity, but understanding what magnetism used to mean is more complicated than simply substituting one word for another. Entering the mental cosmology of Newton, Milton or Kircher entails temporarily abandoning modern beliefs not just about physics, but also about how people interact with each other and the rest of the universe. Three hundred years ago, the word 'magnetic' carried all sorts of emotional overtones and often meant something closer to 'empathetic', a term not actually coined until the 20th century. The nearest approximation then was 'sympathetic': Daniel Defoe, author of *Robinson Crusoe*, described how a magnet draws 'Iron to its self by the Power of Sympathy, or the natural Disposition it has to Embrace that particular Metal'.[3]

Kircher's frontispiece, reproduced as Figure 2, illustrates symbolically how many natural philosophers envisaged a harmonic, hermetic universe bound together magnetically by divine power. At the top of the picture, God's hand pokes through the clouds, holding strong chains to bind together the animate and the inanimate components of His creation. Kircher borrowed this image from Plato, who had pictured poets receiving divine inspiration along a chain of magnetic rings uniting them with God: in 18th-century England, a popular magnetic conjuring trick was called Plato's rings.

Of the three medallions linked together by the divine chains, only the compass needle on the left has obvious connections with modern magnetism. The emblem on the right refers to the concept of sympathetic growth: because plants are magnetically interconnected in different parts of the world, palm trees are drawn towards each other and flourish at the same time. This sympathetic bonding also explained another strange phenomenon – European grape vines apparently burst into bloom simultaneously with the fermentation of wine stored in England. The third medal shows a stag and a cockerel, two creatures said to display instinctive sympathies enabling them to paralyse their prey or escape from their hunters. Just as 'Loadstone draws Iron to it,' one man wrote, so too 'a Ratlesnake fixing his Eyes upon a Squeril, will make him run into his Mouth'.[4]

Figure 2: Magnetic universe. Athanasius Kircher, *Magneticum Naturæ Regnum*, 1667, frontispiece. (British Library)

Kircher was fascinated by the subterranean world because of the extraordinary series of volcanic eruptions in Italy during the 1630s. His magnetic frontispiece extends down below the earth's surface, and reveals part of Kircher's imaginary underground interior, which was riddled with caves, springs and hidden channels occupied by strange lizards such as the ones burrowing their way across the page. Two semicircles show the sunlit hours of the day and the alternating phases of the moon: in Kircher's cosmology, the sun, moon and earth were magnetically bound together not only with each other, but also with plants and animals. On the left, the three-headed sunflower indicates how real sunflowers sympathetically rotate like a clock hand to track the passage of the sun across the sky.

Kircher has followed convention by placing the hot dry masculine sun to the left of his picture, and the cold wet feminine moon to the right. Magnets were often associated with sex. Gilbert had used an explicitly sexual vocabulary of male and female poles coming together in coition, and it is no coincidence that the French word *aimant* can mean either magnet or lover. Metaphorical and physical symbolism were closely twined together. The Romans put powdered magnet into love potions, and Newton's signet ring was mounted with a large magnetic chip, suggesting that he had exchanged it with an exceptionally close friend to guarantee their fidelity to one another. Less privileged lovers who had been deserted tried to draw back the absconder magnetically; suspicious spouses placed a magnet on the bedpost – adultery was confirmed if the magnet had moved.

Erotic magnets provided ample scope for soft porn. Samuel Johnson, the great dictionary compiler, made lewd jokes in his *Rambler* journal about fashioning magnets of different sizes to suit virgins, wives and widows, but magnetic love was also treated more light-heartedly. One of the musicals put on at Marylebone

Gardens was called *The Magnet*, a romantic comedy about a husband's philandering. Once his wife had successfully brought him back home, she sang triumphantly:

> Now to ye, married Fair-ones,
>   Our Counsel is due;
> Of the Magnet be careful,
>   'Twill keep your spouse true.[5]

* * * * *

Because Kircher was a Jesuit from central Europe, he was far more interested in God and the underground world than in the sea. In his frontispiece, a few small sailing boats skim over placid water. But English people lived on an island, and for them sailing across the rough oceans was vital for international trade as well as for warfare. In contrast, Pepys and Tuttell thought magnets were important for navigation and also for profit.

'A treasure of hidden vertues which has made our **Navigation** great, our **Comerce** general, our **Charts**, & **Globes**, much more Accurat & exact.' The lettering on Tuttell's card (Figure 1) is wobbly because each word had to be carved out backwards as mirror-writing before being printed. Nevertheless, the message is clear: magnets may look ordinary, but they conceal a hidden power which had helped to make England the mightiest and richest seafaring nation in the world. Although the Greeks and the Romans never realised it, magnetised needles not only attract and repel pieces of iron – they also point towards the north. In Europe and China, compasses were developed for long voyages which often lasted many months or even years.

Lying across the lower right-hand corner of the picture is a large anchor to symbolise this maritime magnetism. Just above it – next to the cherub's knee – sits a traditional naval magnet: a magnet's power was often advertised by showing it lifting an iron anchor. Before the middle of the 18th century, there were no steel magnets; instead, people used lumps of natural loadstone, an iron ore that was grouped into five different qualities. The finest was a rich deep blue and came from Ethiopia (although the Greeks preferred Magnesia, now western Turkey, where shepherds were apparently magnetically pinned to the ground by their hobnail boots). Navigators could often afford only lower quality reddish stones to re-magnetise their compass needles every few months on overseas voyages. There are several loadstones in this picture, and they have been strengthened by being mounted in iron, a technique called casing or armouring; the prominent suit of armour at the front is a typical Enlightenment visual pun.

On the left of the card is a more realistic test than an anchor of magnetic strength – an armoured loadstone is picking up an iron door key. Even at the Royal Society, natural philosophers often recorded a magnet's power by the number of keys it could support, hardly an accurate measure. The man with the pickaxe is probably digging out some natural loadstone, although his presence might also refer to the way compasses were used to find the way underground in coal mines or to bore tunnels for explosives underneath enemy fortifications. The scene is dominated by Mahomet's tomb, often said to be made of iron and suspended in mid-air by a powerful loadstone.

English people were fascinated by the mighty Muslim empire, and this was a well-known legend that also appeared in encyclopaedias and poems. In this picture, the loadstone seems to be operating directly on Mahomet's body, reinforcing the mysteriousness of magnets as well as of Eastern prophets.

Tuttell emphasised the value of magnets because he was trying to sell them. Can advertising jargon on a flimsy playing card really convey reliable information for historians? Yes, it can – the same messages appeared again and again throughout the Enlightenment. Even small children were taught about loadstone's importance for making Britain great. One reading primer included a fable about an argument between a beautiful diamond and a rusty loadstone. The moral of the story was that outward appearances can be deceptive: although the dirty magnet looks useless next to the beautiful diamond, its hidden powers are worth far more than the jewel's flashy exterior. Educational theories of the time maintained that children learn to read more quickly without being distracted by punctuation, and the talking loadstone boasted breathlessly about its concealed qualities:

It is owing to me that the distant parts of the world are known and accessible to one another that the remotest nations are connected and all in a manner united into one common society that by a mutual intercourse they relieve one another's wants and enjoy the several blessings peculiar to each. Great Britain is indebted to me for her wealth her splendor and her power the arts and sciences are in a great measure obliged to me for their late improvements and their continual increase.[6]

* * * * *

Magnets meant money as well as mystery. *Fatal Attraction* focuses on three scientific investigators, fallible heroes who were lured by nature's strangest power and believed that they could turn it to their own advantage. The first, the astronomer Edmond Halley, tried to map the earth's magnetic patterns and show how the Royal Society could boost Britain's overseas trade and imperial expansion. In the middle of the 18th century, a physician called Gowin Knight sought to make his fortune by selling artificial steel magnets and drastically modernising compass design. And shortly before the French Revolution, Franz Mesmer – who was also a doctor – canvassed political support for his radical medical therapy based on harnessing invisible streams of magnetic fluid.

All three achieved enormous success and were famous in their own lifetime, although they were also publicly mocked by their critics. Their magnetic enterprises have been unjustly forgotten, partly because magnetism itself was such a short-lived science. These men helped to wrest magnetic expertise away from navigators and make it part of natural philosophy. But no sooner had Enlightenment philosophers succeeded in creating a new science of magnetism than it disappeared again, swallowed up into electromagnetism and remembered only dimly as the antecedent of the great electrical revolution in the Victorian era. *Fatal Attraction* tells the stories of these three men as well as many other forgotten makers of Enlightenment magnetism.

PART ONE
HALLEY'S HOLISTIC HYPOTHESES

But the greatest Curiosity, upon which the Fate of the Island depends, is a Load-stone of a prodigious size, in shape resembling a Weavers Shuttle ... This Magnet is sustained by a very strong Axle of Adamant passing through its middle, upon which it plays, and is poized so exactly that the weakest Hand can turn it ... Upon placing the Magnet erect with its attracting End towards the Earth, the Island descends; but when the repelling Extremity points downwards, the Island mounts directly upwards.

Jonathan Swift, *Gulliver's Travels*, 1726

Isaac Newton is commemorated as the great genius of the 17th century, but he was not working in isolation. He relied on the support and criticisms of many other researchers, who were well known to their contemporaries and made their own enormous contributions to science. These vitally important innovators included not only Robert Hooke, Newton's most vociferous rival, but also Edmond Halley (1656–1742), one of his most ardent admirers.

Far more adventurous than the secretive, sedentary author of gravity, Halley was a man of action as well as a man of thought. A brilliant mathematician who became one of the world's leading astronomers and introduced statistics into the shady world of life insurance, Halley also captained an Admiralty ship to South America, tested his own deep-sea diving bells and spied out foreign coasts for Queen Anne. Along the way, he deciphered Arabic manuscripts, wrote Latin poems and ran the Mint at Chester. Hardly surprising, then, that Peter the Great invited him to dinner for free advice about improving the Russian Navy.

Although Halley established his own dazzling career, his achievements have – like Hooke's – been eclipsed by Newton's. He might have been surprised to learn that today he is most famous for a comet that was discovered by neither Newton nor Halley. What is now known as Halley's comet was first spotted in Maryland on 24 August 1682. A fortnight later, Halley observed it from his house in Islington, although 23 years went by before he announced the calculations he had made of its orbit. After sorting out a mistake Newton had made earlier, Halley forecast – correctly, as it turned out – that the comet would return in late 1758 or early 1759. But by then, Halley himself was dead, and so he never knew for certain that his comet would turn up on schedule.

Halley would probably be gratified to realise that his second major claim to fame is his contribution to Newton's *Principia*: Halley is celebrated as the enthusiastic younger colleague who coaxed Newton into finishing his work on gravity, even paying the publishing costs himself when the Royal Society reneged on its initial offer to finance the book. When Halley's comet returned as predicted, these two aspects of his reputation were tied together. Especially in France, the comet's punctuality was an extremely important test of Newton's theories. Its reappearance at the right time helped to convince French mathematicians that the Englishman was right and that their own national hero, René Descartes, was wrong.

Although Halley set himself up as Newton's champion, he did also indicate how he himself might prefer to be remembered – as a magnetic expert. When he was 80 years old, he sat for his last portrait (reproduced as Figure 3). Presumably he attached great significance to the diagram he is holding in his right hand. It illustrated a paper he had published over 40 years earlier in the Royal Society's *Philosophical Transactions* about the internal magnetic structure of the earth. Throughout his life, and

Figure 3: Portrait of Edmond Halley. Michael Dahl, 1736. (Royal Society, London)

well beyond that into the 19th century, Halley was famous for his theories about the world's magnetic patterns and how they affected ships' compasses.

The earth's magnetism might not seem cutting-edge research today, but over 300 years ago, when Halley was first struggling to establish his reputation, it was an opportunistic topic to study. Newton would show that one simple mathematical formula described attraction throughout the universe, and Halley hoped he could work out a general law to account for the mysterious differences in magnetic power at various places on the earth's surface. Even better: magnetism promised to be profitable not only scientifically, but also financially. If Halley could explain how ships' compasses were being affected as they voyaged around the world, then he could market the information to the Navy.

Scientific invention, imperial expansion and commercial gain went hand-in-hand. When a bankrupt collector sold off the contents of his private London museum, his promotional literature glorified the country's international greatness:

> Thus Britain's white sails shall be kept unfurl'd,
> And our commerce extend, as our thunders are hurl'd,
> Till the Empress of Science is Queen of the World ...[1]

If Britannia was to rule the world, then Halley's great ambition was to become the nation's Emperor of Science, the scientific monarch who could predict when comets would return, explain why the earth is magnetic, and show British sailors how to navigate safely round the globe.

· CHAPTER 1 ·
SCIENTIFIC STARTERS

The magnet's potent spell attracts the ore,
Whose strong affinity obeys its power;
Possess'd, diffus'd, its laws impress'd exacts;
The needle points where'er its power directs ...
To God Omnipotent! whose gracious word,
Created all! – and saw that all was good!

Margaret Bryan, *Lectures on Natural Philosophy*, 1806

Fashions change. Although historians used to celebrate the achievements of great scientists without going into details of their personal lives, that approach now seems not only boring, but also wrong. Ambitions and disappointments, political and religious allegiances, financial and emotional problems, all surely affect the sort of science that somebody does. And for people who lived back in the 17th and 18th centuries, it makes even less sense to separate their science from their other activities. Because there were no laboratories into which men could disappear every day and return home only at night, daily life and scientific research were tangled together. Understanding how Halley became Europe's greatest magnetic expert means learning far more about him than the mere factual details of his work with compasses.

But where should a scientific biography start? And how many details are relevant? Researchers are understandably reluctant to cut out a sentence that represents a week's research in an archive, but should every little snippet of information be included? Unfortunately, only too often there are few choices to be made, because there is so little material left to work with. Halley kept no diary, and very few personal letters survive, so it is hard to penetrate his persona. Unlike his friends Pepys and Hooke, who wrote such intriguing accounts of their most intimate experiences, Halley left behind him virtually no information about his wife, his children or how he spent his spare time, let alone juicy details of his sex life.

Biographers must face up to other problems as well. When children are born, there is no way of predicting that they will be famous in the future (unless, of course, they are part of a very special family such as royalty). However much Halley's parents doted on their oldest son, they could never have forecast that his name would be emblazoned across the newspapers every 70 or 80 years when comet 1P/1682 Q1 reappears in the sky. But it is impossible for us to put ourselves into their state of ignorance, because their future is already our past. Halley's story is fascinating only because we already know that it has a happy ending – he *will* predict that comet's orbit correctly.

One way of resolving these difficulties is to abandon any attempt to describe someone's life in the order in which it was lived: instead of writing chronologically, biographers write thematically. From a modern perspective, Halley's achievements can be divided into chunks which fit neatly into separate chapters. His astronomy, magnetism and maps seem to be distinct not only from each other, but also from the other leitmotifs of his life – his alleged atheism, his zest for travel, his interest in Arabic manuscripts.

But when describing a man of the Enlightenment, using that sort of compartmentalisation does not make sense. Take religion. For one thing, claiming to follow the right religion was important for a man who wanted to have any sort of academic career, since only adherents to the Church of England could go to university: Jews, Catholics and other Christian groups such as Quakers were excluded. So it was essential for Halley to present himself outwardly as an orthodox believer, although he failed to get at least one job – a professorship of astronomy at Oxford – because his interviewers doubted his piety.

Christianity and natural philosophy were inextricably bound together at a more fundamental level. Because many people still saw the Bible as the true source of knowledge, Halley had to make any new theory he introduced compatible with biblical explanations. He explicitly fashioned some of his magnetic ideas to display his religious conformity. Similarly, his fascination with Babylon and other ancient cities was not merely an antiquarian hobby. Through studying Arabic texts and their Latin translations, Halley concluded (and his conjectures are confirmed by modern astronomers) that the moon is speeding up as it orbits around the earth.

Fatal Attraction is a book about Enlightenment magnetism, neither a biography of Halley nor a compendium of his impressive and varied achievements. However, understanding his commitment to solving the world's magnetic puzzles entails tracing what else happened in his life. What special appeal did magnetism hold for Halley? How did he embark on this research, and how did it relate to his other activities? One place to begin this magnetic tale is at Halley's birth. But be warned – this is not a traditional cradle-to-grave account of Halley's life.

* * * * *

Pinpointing Halley's own beginning is difficult. Many church records were destroyed during the Great Fire of London in 1666, so no birth certificate for him survives. The best bet seems to be his own statement – 29 October 1656 – which also appears on an astrological horoscope somehow preserved over the centuries in Oxford. Halley's father was a wealthy property owner who decided to give his clever son a good education. He was sent first to St Paul's in London, and then to Queen's College in Oxford.

Although we have no first-hand comments about Halley's youthful personality, it is obvious that he was a brilliant mathematician and was obsessed by science. Even as a schoolboy, he made astronomical and magnetic measurements which came in useful for articles he wrote later. He was only twenty years old, still an undergraduate, when his first three papers appeared in the *Philosophical Transactions*: a discussion (in Latin) of a mathematical method for working out the elliptical orbits of planets from astronomical observations, a report on sunspots, and calculations of the time differences between Oxford, Greenwich and Danzig (now Gdansk).

Halley had also already learnt how to promote himself. Only a week after the king had appointed his first official astronomer, John Flamsteed, Halley sent this new Astronomer Royal a letter boasting about the accurate measurements he could make with his own instruments. His strategy worked. Soon Halley was travelling backwards and forwards between London and Oxford, setting himself up as Flamsteed's indispensable assistant and making invaluable contacts.

All the networking and astronomical dedication paid off. A year later, the king sent a letter to the East India Company asking them (well, ordering them) to carry Halley to St Helena, a small island in the Atlantic roughly level with modern Brasilia. Halley had rejected mainland Brazil because he thought it would be a waste of time to learn the language. St Helena was British and had nice weather: not a flippant reason, but an important consideration for an astronomer. From his vantage point on this island lying far to the south, Halley planned to map the positions of southern stars invisible from the northern hemisphere. He also wanted to observe a Transit of Mercury, when Mercury passes in front of the sun. These Transits occur only about thirteen times a century, and they help astronomers to calculate the dimensions of the solar system.

The ship took three months to get to St Helena, but Halley made only one laconic comment about the voyage – *satis foelicem*, pleasant enough. So how could a twenty-year-old pass the time? To start with, he put into practice all the theoretical methods of navigation he had studied at school and university. The sun, moon and stars were a ship's most vital guides, and Halley presumably found out how to carry out all the necessary measurements and mathematical observations far more efficiently. What else could he do? Because of pirates, storms and uncharted rocks, travelling by sea was very dangerous during the Enlightenment. So on the way to and from St Helena Halley must have thought about ways of helping sailors (and himself) to survive such hazardous journeys.

Magnetic navigation was certainly an area that needed reform, and he later commented on some magnetic measurements he had made near the equator.

Before he set off, Halley made sure that he was equipped with the most up-to-date astronomical instruments available. In contrast, the ship's navigators relied on traditional techniques that had been developed over the centuries. Ships did carry several magnetic compasses, but they were generally old, inaccurate and poorly maintained because the sailors knew that they were of only limited value. The compasses' main purpose was to indicate the general direction in which to steer: seasoned mariners were happy to accept readings accurate to around ten degrees.

One pair of steering compasses was used by the helmsman. These were kept in a small cupboard called a binnacle, which was divided into three compartments – a central one for a candle at night, and one on either side to hold a steering compass so that the helmsman could easily take a reading wherever he was standing. Another type of compass, sometimes called a crown compass because of the small crown included in the design, now looks strange because the west and the east are the wrong way round, as if reversed by a mirror. This is because these compasses were made to hang upside down from the ceilings of officers' cabins, so that men could check the ship's course without having to get out of bed.

Magnetic compasses were sold by many instrument makers besides Thomas Tuttell. Figure 4 shows the advertising card of John Bennett, who set up shop in Soho, then to the west of London rather than the centre as now: London later expanded westwards against the

Figure 4: John Bennett's trade card, c. 1760. (Science Museum, London)

prevailing winds, because wealthy citizens wanted to avoid the stench and disease of the dockland area to the east. Bennett was trading in the middle of the 18th century, but Halley would have found the images he chose for his card familiar; like his competitors, Bennett showed out-of-date instruments because he was reluctant to pay for new printing plates to be engraved.

The term 'scientific instrument' was introduced only in the 19th century. Before then, instruments were – as Bennett's card shows – divided into three categories: mathematical, philosophical and optical. This is because, as new devices were developed, existing craftsmen expanded their traditional field of expertise to include them. Instrument shops were passed down through the generations, as sons and daughters gradually took over their parents' work. Women in these skilled artisan families often played strong managerial roles, and many of them also made instruments. Even Flamsteed trained his own wife, Margaret, alongside his other apprentices; her surviving notebooks show that she was an accomplished mathematician, and she played a crucial role in publishing the Astronomer Royal's observations after his death.

Opticians who had previously specialised in making magnifying glasses and reading spectacles (like the ones illustrated at the bottom centre of this card) diversified by learning how to make telescopes and microscopes. Philosophical instruments included the newer pieces of equipment invented by natural philosophers – airpumps, thermometers, barometers, electrical machines. Compasses and magnets were classified not with the philosophical instruments, but with the mathematical ones. This meant that they were grouped with geometrical drawing instruments such as dividers and protractors, and also with geographical tools such as sundials (bottom right of the card) and globes (the one in the centre here shows the earth's surface, but would have been sold in a pair with a celestial globe of the heavens). Still more importantly, this category of mathematical instruments included the day-to-day tools used by practical mathematicians – chains, rules and theodolites for surveyors, callipers for gunners, and octants, compasses and hourglasses for navigators.

The top half of Bennett's card is dedicated to seafaring. On the top left, an astronomer peers through the eyepiece of his octant, an angle-measuring device which incorporated a rotating mirror and other improvements suggested by Newton and Hooke; its most common use was for working out a ship's latitude by sighting the pole star. Hanging down beneath this navigator is another of his essential instruments – a portable telescope of the type taken on board ship. And on the right is a cylindrical armed block of loadstone, proclaiming its strength by supporting a ship's anchor.

The most prominent instrument is the one surrounded by rococo curlicues at the upper centre – a magnetic azimuth compass. This resembles an ordinary steering compass, but has a measuring device mounted on top of it for taking bearings from the sun or the stars. Compasses like this were designed to overcome one of the major obstacles to magnetic navigation – magnetic variation. This term is still used in navies and air forces, although scientists now call it declination. Experienced mariners knew that the needle of a compass does not point directly towards the earth's geographical north; instead, it settles in a direction to one side of the northward direction. The angle between the geographical north and the needle is called the angle of variation. To make matters even worse from a sailor's point of view, this angle of variation itself varies, depending on which part of the world his ship is in. The needle may point either to the west or the east of true north, ranging from 0° along a line running roughly diagonally across the Atlantic, to very high values near the north and south poles. And there is a further complication, one that had been discovered only far more recently, in 1634: the patterns of the earth's magnetic variation change over time. By checking measurements over the past hundred years, Halley's contemporaries realised that the variation had changed by over 15° in London.

When Christopher Columbus was sailing to America, he knew that this problem of magnetic variation existed, although he thought that it remained constant with time. The great Portuguese navigators of the 16th century invented ways to measure it, and by the time that Halley sailed to St Helena, seamen had devised several ways of overcoming the difficulty. For voyages near the European coast, where the variation is fairly uniform, it was easy to compensate for the discrepancy by permanently rotating the dial (face or card) of a steering compass beneath the needle, so that it appeared (falsely) to be pointing towards the geographic north. For longer journeys across the Atlantic, an azimuth compass could be used to correct the readings on the steering compasses.

The basic principle of the azimuth compass was straightforward. For the model shown on Bennett's card, a navigator was meant to line up the instrument with the sun, whose position was already known (for example, at noon it lies due south in the northern hemisphere). Then he should look at the shadow cast by the diagonal string on to the dial of the compass, and compare it with the direction of the compass needle. In practice, life was not so simple. For one thing, some nifty trigonometry was needed to transform the readings into the magnetic variation. Moreover, even during the daytime the sun is often not shining in the middle of the Atlantic, and at night, stars had to be used instead. And in addition, compasses were intrinsically inaccurate, especially since measurements that were relatively easy to carry out on land seemed far more challenging in a heaving boat during a storm.

So with all these complications, why should Halley (or anyone else) have bothered to think about magnetic variation? One answer to that question was – longitude. When ships sail across oceans, navigators need to be able to work out their longitude accurately so that they know where they are and where to find the next safe spot to land or the nearest dangerous reef to avoid. For countries like Portugal and Britain that were expanding their international trade, it was vital to find a method for measuring longitude at sea. The first financial reward was offered by the Spanish king in 1567; 30 years later, his successor increased the prize, and rulers in Holland, Portugal and Venice also set up their own schemes.

Eager inventors, scathingly referred to as projectors, proposed many ingenious plans, and finding the longitude became a standard joke: like squaring the circle, it seemed an impossible task. When the Royal Society was being set up in the early 1660s, one satirist wrote a long poem mocking its aims. Verse 26 (out of 28) ran:

> The Colledge will the whole world measure,
> Which most impossible conclude,
> And Navigation make a pleasure
> By finding out the longitude.
> Every Tarpalling shall then with ease
> Sayle any ship to th'Antipodes.[2]

(The 'Colledge' was Gresham College, where the founders of the Royal Society had been meeting; 'tarpalling' referred to a tarpaulin, or tar, slang for a common sailor; over a hundred years before Cook landed in Australia, the Antipodes seemed distant and almost mythical.)

In principle, suggestions relying on magnetic variation did seem feasible: if you know in advance what the variation is in several different places, then you can use an azimuth compass to measure the variation and hence find out where you are. The very first volume of the Royal Society's *Philosophical Transactions* included '*Directions for Sea-men going into the East & West Indies*', because the new Society wanted to enlist navigators as unpaid data collectors to help solve the longitude problem. They were asked to record all sorts of information, but the first requirement was to make magnetic readings at different places as they sailed around the world. They were then supposed to make three copies of all their measurements (this was long before carbon paper had been invented); surprisingly, some travellers obeyed these time-consuming instructions.

* * * * *

Halley did make some magnetic observations during the year he spent on St Helena, but at this stage he was far more committed to measuring the stars: he hoped that astronomy would both solve the longitude problem and also establish his reputation. The weather was atrocious, very different from the clear skies he had been promised, and Halley was also dismayed by the obstructive behaviour of the island's governor (he was eventually fired, although the papers detailing his misdemeanours are oddly missing from the files). Nevertheless, Halley dedicated himself to observing stars, eclipses and nebulae. Unlike Flamsteed, who became embroiled in vitriolic arguments about who had compiled some data first, Halley already appreciated the importance of getting into print quickly. Within a few months of his return to England, he had published his comprehensive *Catalogue of the Southern Stars*, been elected a Fellow of the Royal Society, and resumed his cultivation of influential men. Halley was also rich – his allowance from his father was three times the salary of the Astronomer Royal.

The next year Halley was off again, this time to Danzig. There he stayed with Johannes Hevelius and his wife Elisabetha, whose rooftop observatory by the Baltic boasted some of the largest instruments in Europe. Halley was engaged on an expedition of scientific espionage. His English backers, Hooke and Flamsteed, believed that the telescopic sights they had invented gave the best results, but Hevelius disagreed – he was nicknamed the lynx of Danzig because he insisted on lining up his giant brass sextant with the naked eye, without the help of additional lenses. Halley's mission was to discover how Hevelius managed to get extremely accurate readings.

Installed right inside the enemy's camp, Halley knew that he had to be tactful, but he seems to have slightly overdone the charm. Flamsteed became furious with Halley for writing flattering testimonials to Hevelius, and gossip also circulated about Halley's night-time observation sessions with Elisabetha Hevelius. Why, people wondered, had he lingered far longer than anticipated in Danzig, apparently too busy even to send letters back to England? Years later, Halley's enemies were still insinuating that he had made Johannes Hevelius 'a Cuckold, by lying with his wife when he was at Dantzick, the said Hevelius having a very pretty Woman to his Wife, who had a very great Kindness for Mr Halley and was (it seemed) observed often to be familiar with him'.[3]

After he returned from Danzig, Halley lived at home with his father for a year, and then set off for a trip round France and Italy. His was an unconventional Grand Tour. Whereas most wealthy young men in their early 20s claimed that they were visiting Rome in order to gratify their interest in ancient statues, Halley declared that his obsession was astronomy. He was still on the road to Paris when he witnessed a stunning phenomenon – the return from behind the sun of the great comet of 1680. His letters to Flamsteed and other astronomical colleagues were full of excitement; like them, Halley believed that comets are governed magnetically. This particular comet was exceptionally bright and was visible for four months. Although it fascinated natural philosophers – especially Newton – it terrified many other people, who declared that this extraordinary fire in the sky could only presage disaster.

As he travelled through Europe, Halley kept a very sober travel diary. He made careful listings of all his astronomical measurements (and some magnetic ones), although he complained to Hevelius about the lack of local enthusiasm for astronomy in Rome. He consoled himself by touring the ancient ruins not only to admire them, but also to compare the lengths of Roman, Greek and English feet, an analysis of changing measurements which was important for assessing the reliability of old maps.

Three months after he arrived back in England, Halley got married and moved to Islington, where he embarked on a long study of the moon and the planets. His goal was longitude: if he could correct existing tables showing the moon's motion, then sailors at sea would have a more reliable way of working out their longitude. Halley had aimed to continue this work for an eighteen year cycle, until well into his 40s, but his plans were abruptly interrupted after only a couple of years when his father lost his fortune and died under mysterious circumstances. But before that, during this tranquil interlude on the northern outskirts of London with his new wife Mary, Halley published his first article on magnetism.

CHAPTER 2

MAGNETIC VISIONS

Change, as ye list, ye winds! my heart shall be
The faithful compass that still points to thee.
John Gay, 'Sweet William's Farewell to Black-eyed
Susan', 1720

'*A theory of the* Variation *of the* Magnetical COMPASS': by contemporary standards, Halley chose an unusually short title for his first paper on magnetism. Writing with great panache, he used the same tactic of self-promotion as he had for some of his earlier articles on astronomy – he attacked his predecessors. Most writers, he declared, simply fail to comment on the important phenomenon of variation; and even those who do mention it, he continued, offer superficial theories that are not worth taking any further. As well as criticising Gilbert, Kircher and Descartes, Halley scoffed at Henry Bond, a lecturer in navigation who had been discussing the problem for the previous 40 years. Bond had recently published a small book, *The Longitude Found* (soon cruelly rebutted by a critic who wrote *The Longitude Not Found*), but Halley dismissed Bond's work because it was restricted to London. Halley himself would, he announced, formulate a hypothesis for the entire world.

* * * * *

Halley was still in his 20s when he told the Fellows of the Royal Society about his revolutionary magnetic ideas.

Soon they were published in what was then the world's most prestigious scientific journal, the *Philosophical Transactions* of London's Royal Society. With very few competitors, the *Transactions* enabled researchers all over Europe to keep up-to-date with the latest developments.

Halley carefully advertised the significance of his research by emphasising that it would be practically as well as theoretically valuable. To do this, he staked out an ambitious claim that reads like a manifesto for Enlightenment research into terrestrial magnetism:

> ... the great utility that a perfect knowledge of the Theory of the Magnetical direction would afford to mankind in general, and especially to those concerned in Sea affairs, seem[s] a sufficient incitement to all Philosophical and Mathematical heads, to take under serious consideration the several *Phænomena*, and to endeavour to reconcile them by some general rule ...[4]

Studying variation was, Halley explained, essential to rescue 'one of the noblest Inventions' that had ever been made – the navigational compass This was not just Halley's hype. The philosopher John Locke declared that 'He that first discovered the use of the compass did more for the supplying and increase of useful commodities, than those who built workhouses [workshops, early factories]'.[5] Locke belonged to the modernist school of thought, which emphasised the importance of three major Renaissance innovations. The other two were the printing press and gunpowder – and according to enthusiasts, this triplet of inventions had made

European civilisation even greater than that of the ancient Greeks.

Campaigners for progress and commercial expansion continued to celebrate magnetic compasses right through the 18th century. With a novelist's nice turn of phrase, Defoe rejoiced in the explosion of trade and exploration that they had unleashed in the 15th century: 'Navigation being as it were let loose, and the Seaman's Hands unty'd, which were fetter'd and manacl'd before by their Ignorance, not daring to venture far from the Shores; I say in consequence of this great Discovery, all the *European* Nations went to work, spreading the Seas with Ships, and searching every part of the Ocean for new Worlds.'[6]

Halley knew that stressing the commercial importance of his work was one of the best ways to hold the attention of natural philosophers. In his ambitious policy statement, he also expressed a second great goal of Enlightenment optimists. For them, understanding the world meant imposing order onto unruly phenomena, and Halley aimed to unify magnetism's disconcerting fluctuations into one single pattern. In modern terms, he promised to deliver a Grand Unified Theory of Magnetism (although he immediately protected himself against criticism by admitting that he did not have enough observations).

But first his readers had to plough through several pages of figures, problems, and more denouncements of his rivals. Halley was right to admit that he did not have many observations – he had a total of only 55 from around the world, which included some of his own as well as others going back to 1580, when instruments were even less reliable. The Spaniards and the Dutch had proved uncooperative but, Halley reassured his readers, the measurements he did have were 'mostly' made by skilled people. And when concrete evidence was missing, as for the north Pacific route to Japan, he helpfully suggested what the figures should be according to his hypothesis.

Enlightenment philosophers loved words, and they typically gave long verbal descriptions of data which would nowadays be summarised as tables or graphs. Being able to interpret a diagram involves understanding a kind of visual language, and although that is second nature for modern scientists, Halley and his colleagues had not developed the necessary conventions. Although Halley became one of the great visual innovators of the Enlightenment, in this early paper he followed custom by wordily discussing the complexities of variation as he travelled on an imaginary magnetic trip around the globe.

Emphasising confusion and sneering at earlier suggestions are both good rhetorical strategies for making a new solution seem attractive, and Halley divulged his own hypothesis only after several pages. Although vague on details, his basic proposal was that the entire earth is one great magnet. So far, not so very different from Gilbert, who used a small sphere of solid loadstone (called a terrella, or miniature earth) to model the world's magnetic characteristics. However, Halley went much further by suggesting that this giant terrestrial magnet has not two poles, but four – two in the northern hemisphere and two in the southern – which are of different strengths, distributed unevenly around the globe, and mysteriously wander about over the years. He suggested that a compass needle would be affected most powerfully by the pole nearest to it. For example, he could explain the line of zero variation running diagonally across the Atlantic by assuming that there are two northern magnetic poles, one near northern Europe, and one near northern America, which both tug on a compass needle at the same time, but in different directions.

Halley knew that this was all hypothesis, and he listed three major problems that had to be surmounted to improve the foundations of his theory. First of all, more magnetic measurements needed to be made on land. Because magnetism had been far more important for navigators than for natural philosophers, the seas were far better charted than the continents. Another problem was finding out how the power of a magnet decreases with distance. No reliable data were available to calculate accurately how the effect on a compass needle would diminish as it was carried further away from one of the earth's magnetic poles. And finally, because the variation was itself changing with time, Halley admitted that it would take hundreds of years to establish a complete theory.

Despite these reservations, Halley was so confident that he printed his conclusion in italics. 'I have put it past doubt', he boasted, *'That there are in the Earth Four such Magnetical Points or Poles which occasion the great variety and seeming irregularity which is observed in the variations of the Compass.'* And he ended his paper with a fine flourish. The patterns of magnetic variation are, he declared, 'secrets as yet utterly unknown to Mankind; and are reserved for the Industry of future ages'. However, only eight years later he conducted the Fellows of the Royal Society on a second magnetic mystery tour.

* * * * *

A lot happened in those eight years. Most importantly, Halley grew close to Newton, who was fourteen years older than him. It was Halley who recognised the importance of Newton's mathematical work on the planets, Halley who persuaded him to write a book, and Halley who paid for Newton's *Principia* to be published. Halley also nurtured his own career by taking on the time-consuming job of Clerk at the Royal Society on top of his many other activities. At home he was trying to sort out the intrigues surrounding his father's death, as well as coping with the birth of at least two babies (there are virtually no records, although two daughters did survive). More publicly, he was charting the Thames, inventing a diving bell, mapping the winds, writing a long Latin poem for the *Principia* ... In addition, unpublished letters and manuscripts reveal that Halley was still thinking about magnetism.

Halley was also acquiring a new enemy – Flamsteed, his former protector. The conflict between them arose partly because the two astronomers had different opinions about how science should be conducted. Flamsteed was a single-minded observer who guarded his measurements closely, and he later became enmeshed in a vitriolic argument with Newton about who owned the data he had collected. In contrast, Halley thought globally. He turned his attention to an enormous range of topics and freely mixed his own and other people's readings in order to establish large-scale patterns of phenomena such as winds, tides and magnetism. Among other complaints, Flamsteed accused Halley of stealing his four-pole hypothesis from Peter Perkins, the head of the Mathematical School at Christ's College. Perkins was already dead, but although Halley probably did use Perkins's observations, he regarded that as a legitimate tactic in scientific research.

Halley's key visit to Newton took place in August 1684 at Cambridge, when Halley discovered how much progress Newton had made on an astronomical problem that was also troubling natural philosophers in London. Hooke, Halley and Christopher Wren realised that Kepler's laws describing how the planets move magnetically around the sun can be explained by assuming that the sun attracts each planet with a force varying as the inverse square of the distance (so that doubling the distance means that the force becomes four times weaker, tripling it means nine times weaker, and so on; this inverse-square law is the fundamental relationship at the heart of Newton's work on gravity). The difficulty lay in proving it mathematically. The next time Halley saw Newton, he made him promise to organise his ideas into a book – and also to write it quickly. Newton took his advice to heart, and did little else for the next eighteen months.

Back in London, Halley embarked on some preliminary publicity. Summoning up the strategy which had worked so well for him in the past, in 1686 he published a paper on gravity which started by systematically attacking every existing interpretation. René Descartes, Christian Huygens, Isaac Vossius – Newton's great predecessors had all got it wrong, Halley claimed. Only the previous year, Halley had written (in Latin) to one of his European colleagues explaining that magnetism and gravity did not seem so very different. But now he was scathing about people who linked them together. 'Some think to Illustrate this *Descent* of *Heavy Bodies*, by comparing it with the Vertue of the *Loadstone*', he sneered; but 'this Comparison avails no more than to explain *ignotum per aeque ignotum* [the unknown by the equally unknown]'.[7]

While he was tactfully coaxing Newton into completing the *Principia*, Halley continued with his magnetic research. As he had pointed out in his article, in order to build a good model of terrestrial magnetism he needed to know how a magnet's attractive power falls off with distance. Criticising Hooke's attempts, Halley set up his own experiment outside in the courtyard of Gresham College, so that his loadstones and compass needles would not be affected by the iron objects present inside a building. Yet despite these precautions, his results were inconclusive. Neither Halley nor Hooke ever did derive a mathematical law – unlike some of their successors, who managed to persuade themselves that they had found an inverse square law to match the one governing gravity.

Nevertheless, Halley did come up with a new hypothesis about the earth's magnetism. Four years after the *Principia* finally appeared, he stood up before the Fellows of the Royal Society and told them about his new version of hypothetical wandering poles. When this second magnetic article was later published with a few alterations, Halley dominated that issue of the *Philosophical Transactions*, the Royal Society's journal. It contained only three papers, and two of them were by him – a short mathematical one on infinite quantities, and his far longer one on magnetism. Sandwiched in between was somebody else's discussion of trumpets. Because paper and printing were expensive, illustrations were often squashed close together, and so musical notation surrounds Halley's magnetic diagram, reproduced here as Figure 5 (and the same as the one he is displaying in his portrait, Figure 3).

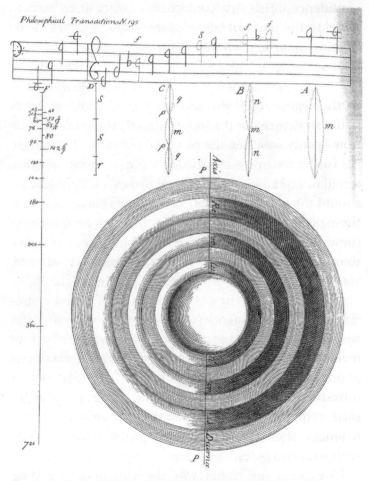

Figure 5: Edmond Halley's diagram of the earth's magnetic interior. *Philosophical Transactions* 16, 1692, facing p. 555. (Cambridge University Library)

Despite the cost of producing this picture, Halley referred to it only right at the end of his paper. He had devised two models of the earth's interior structure, and he spent most of his time persuading his readers to accept the less outlandish one. Only after he had won their confidence in his first conjecture – which itself seemed novel to the point of being bizarre – did he conclude by offering them an even stranger one.

* * * * *

At the beginning of this second magnetic paper, Halley had only one major predecessor to criticise – himself. He now openly admitted the two most obvious objections to his earlier four-pole model. All magnets, even spherical terrellas, have only two poles, so why – Halley asked – should the earth have four? And in addition, how could these poles move about inside a solid magnet? To resolve these difficulties, Halley modified his first plan rather than rejecting it outright. Gradual rotation, he decided, was the key to the solution.

Aware that his new hypothesis could well seem 'Extravagant or Romantick', Halley asked his audience to suspend their disbelief and take it seriously. First he reviewed what was known about the complicated fluctuations in the earth's magnetic characteristics, and then he introduced his revised model. Halley proposed that the earth is divided into two separate parts: an outer hollow magnetic shell and an inner revolving magnetic core, each with two poles.

To explain the changes of the variation over time, Halley suggested that this internal nucleus rotates westwards like the earth itself, but at a very slightly slower rate than the outer cortex. This discrepancy arose, he suggested, because when the earth had first been set spinning on its axis, its initial impulse was imperfectly communicated to the central sphere, so that every 24 hours it lagged a tiny amount further behind the outer shell. This structure would, he explained, make it appear as if each hemisphere had one fixed and one wandering magnetic pole.

Halley estimated that it would take 700 years for a full cycle of relative rotation to be completed, far too long for anyone to be certain that his theory was right. But although this conveniently placed him beyond immediate attack, Halley urged travellers to measure magnetic variation all over the world and send in their readings to the Royal Society. Only then, he insisted, could his theory be corroborated. A few years later, Halley embarked on two reconnaissance trips of his own across the Atlantic, but his immediate aim was to pre-empt facile criticisms of his unusual model.

First of all, he made it seem less unusual – well, not unique, at any rate. He invited people to think about Saturn: if a central planet and an outer ring could rotate together in a stable state, he argued, then why not an inner sphere and an outer shell? Then Halley tackled another question – why doesn't the sea leak through into the space between the outer shell and the inner core? Again, he pointed to parallel examples, such as beds of chalk or clay. Arguing by analogy in this way was seen as being far more valid then than now, partly because the ultimate explanation was God's will. Since God had created so many marvellous yet inexplicable natural wonders, Halley reasoned, then surely He could find a way to make the earth's inner surface watertight. In fact, Halley continued, making it magnetic might be God's way of preventing crumbs falling off and being ineluctably attracted down towards the centre by gravity, as had been so recently shown by 'the excellent Mr. *Newton*'.

Halley also called on Newton for a more mathematical vindication of his ideas. One of the major philosophical puzzles at this time was to relate the patterns of the tides to the movement of the moon. In the course of discussing the lunar influence on the oceans, Newton had concluded that the moon is almost twice as dense as the earth, in the ratio of nine to five. Halley now provided a straightforward interpretation of this difference – the moon is solid, whereas the earth has an internal cavity which takes up four-ninths of its volume. This lent further support to Halley's magnetic hypotheses.

Ambitious and self-confident as Halley appeared to be, even he would perhaps not have predicted that this basic model would remain valid well into the 19th century. To construct his hypotheses, Halley had started with a small set of magnetic measurements taken on the earth's surface, and extrapolated downwards to imagine what the world's invisible interior might be like. For most people, the outside of the globe was far more important. Financial rewards were more tempting than intellectual ones, and improving navigation promised the greatest gain for magnetic experts. During the next hundred years, Halley as well as many professional mariners compiled far more detailed and accurate information about the earth's observable magnetic characteristics. The British were often at war with the French and the Spanish, but Halley urged them to collaborate in the name of progress. As he boasted, 'I am in hopes I have laid a sure Foundation for the future Discovery of an Invention, that will be of wonderful Use to Mankind when perfected; I mean that of the Law or Rule by which the said Variations Change, in Appearance regularly, all the World over.'[8]

* * * * *

So far, Halley had made no reference to the diagram that accompanied his article and which he displayed so proudly in his portrait (Figures 5 and 3). Almost as an aside, Halley mentioned that if his basic model failed to explain the variation satisfactorily, then it could be modified by introducing more shells, each with two poles. As he neared the end of his paper, he gradually slipped into this more complicated scheme. He knew that other natural philosophers were also fascinated by the terrestrial secrets concealed beneath the earth's surface. However, the questions they posed did not always resemble those of modern scientists. They wondered whether earthquakes and volcanoes confirmed the ancient notion that hell lies deep inside the globe. And they worried about the biblical account of Noah's flood: where could all that water be except in a giant internal reservoir?

Halley was worried about his career prospects. Now that his father and his fortune had disappeared, he needed a job. When he gave this second talk about magnetism on 25 November 1691, he had already applied for a professorship in astronomy at Oxford and persuaded the Fellows of the Royal Society to write a strong reference for him. But he knew that rumours were circulating about his commitment to Christianity. The influential Bishop of

Worcester was alarmed to hear that Halley was 'a skeptick and a banterer of religion.' He asked his chaplain, Richard Bentley, to interrogate Halley, who obstinately declined to give ingratiating answers. Despite the Royal Society's recommendation, the Oxford post went to one of the other applicants.

If Halley were to find work, he needed to establish his orthodoxy for the future. When the printed version of his lecture appeared a year later, the only one we have available to study, Halley informed his readers that he had spruced up some details of his religious arguments. To contradict his reputation of being a heretic, Halley needed to emphasise that God was central to his magnetic hypotheses. And that was why he dedicated the final section of his paper to a theological discourse on the magnetic mysteries concealed inside the earth.

Halley made his magnetic model support the Christian belief that the universe is not eternal. To imply a finite ending, Halley suggested that an aether – a mysterious invisible cloud of gas spreading throughout the universe – might be gradually slowing down all the heavenly bodies. One major objection to this aether was that it should have swept the moon away to become an independent planet revolving around the sun. Halley triumphantly pointed out that his magnetic scheme resolved that difficulty, because his semi-hollow earth meant that the moon was relatively more dense, and so could be held in an orbit about the earth.[9]

The aether evidently continued to trouble him, since on the publication day itself – 19 October 1692 – Halley raised it again during another talk at the Royal Society. This time he tried to prove its existence by arguing that light must be carried through the universe by some sort of physical medium, however rare that might be. Even twenty years later Halley was still extremely sensitive to accusations that he deviated from Christian orthodoxy by believing in an eternal universe. Halley did eventually acquire the Savilian chair of astronomy at Oxford, but he never stopped worrying about his religious critics. In an article about the saltiness of the seas, he maintained that his results showed that the earth's age was limited: he knew that it was vital for natural philosophers to make their accounts of the past compatible with the version given in the Bible.

Whatever his personal faith, Halley was determined to display an orthodox position in his paper on magnetism. He reserved his strongest religious argument for the end, when at last he presented his more complicated scheme, the one illustrated in his diagram (Figure 5). The shaded portions, Halley explained, represent 'Magnetical Matter'; the earth has an outer crust 500 miles thick, which surrounds three concentric magnetic globes whose diameters are proportional to those of Venus, Mars and finally Mercury, a solid ball 2,000 miles in diameter.

Halley then casually announced what now seems an extraordinary notion for an ambitious natural philosopher trying to establish his credibility. He claimed that the lighter spaces between the shells are inhabited. Although he never specified exactly what sort of creatures might live beneath the earth's surface, Halley suggested that God had provided an atmosphere and a special type of light to support life on these inner planet-like bodies (strangely, although he was very keen to stress God's forethought, he never mentioned food).

Although Halley's suggestion was never widely accepted, it did not – as would be the case now – alienate his audience. On the contrary, Halley was aiming to gain support by bringing God directly into his argument. According to Halley, God had economically provided extra living space by maximising the interior surfaces of the earth. After all, he argued, 'We our selves, in Cities where we are pressed for room, commonly build many Stories over the other, and thereby accommodate a much greater multitude of Inhabitants.'[10]

'God does nothing in vain': this was a common maxim among natural philosophers who wanted to justify some particular odd feature that they had discovered. In England, many of them were keen proponents of natural theology, convinced that studying nature would lead them to a greater understanding of God Himself. Natural theologians excelled in finding people-centred reasons for all of God's creations – He made mountains so that melting snow would give us water to drink, He placed stars in the sky so that human beings would marvel at His splendour, and He gave us horses so that we could ride from place to place. Some of these explanations do now seem over-ingenious. Did God really invent ocean tides to help our ships sail in and out of ports? Did He really place carnivorous animals on the earth so that they would reduce the net amount of suffering by giving their prey quick and merciful deaths?

Like his well-educated colleagues, Halley had been trained in the classical art of rhetoric. Because he presented his arguments so skilfully, it is hard to be sure how firmly he believed in his own multi-shell model. Halley advanced strategically by posing unanswerable questions.

He challenged his readers to think up 'a less absurd' hypothesis, and pointed out that since God had created such a great variety of creatures to live on top of the earth – birds, flies, fishes, reptiles, mammals – then why should we think it strange that He should have placed others underneath? He gave his mysterious underground light a similar treatment. Since other types of luminosity exist that we do not understand, Halley argued, then there is no reason to think that we should be able to solve this problem. Perhaps, he suggested hopefully, the atmosphere between the shells is itself luminous; or perhaps the magnetic arches are coated with some strange substance that radiates light.

Halley made his religious motives clear, almost as though Bentley and his Bishop were listening. 'I have adventured to make these Subterraneous Orbs capable of being inhabited', Halley explained, 'designedly for the sake of those who will be apt to ask *cui bono* [for what good], and with whom Arguments drawn from *Final Causes* [God] prevail much.'[11] But however hard he tried to make himself appear a respectable natural theologian, Halley knew that his vague scheme was very sketchy, and he never referred to it explicitly again. Over time, it seems to have become gradually amalgamated with his simpler model, so that the mysterious light and the subterranean life were located in the single space between the inner magnetic globe and the earth's outer crust. Halley promised a more polished account – but a revised version never did appear.

· CHAPTER 3 ·
INSIDE THE EARTH

For the Life of God is in the Loadstone, and there is a
magnet, which pointeth due EAST …
For due East is the way to Paradise, which man
knoweth not by reason of his fall …
For the Longitude is (nevertheless) attainable by steering
angularly notwithstanding.

Christopher Smart, *Jubilate Agno*, 1758–63

Halley's basic model of the earth's magnetic interior – a
single inner globe rotating inside an outer shell – was
accepted well into the 19th century, and it established his
reputation as England's great magnetic expert. Most
people used his scheme to help them understand the
magnetic patterns on the earth's surface. But a few
researchers were, like Halley, fascinated by the details of
the magnetic mysteries hidden beneath the earth's crust.

One of them was William Whiston, a brilliant mathematician who had been Newton's protégé at Cambridge
before being expelled from the University because he
publicised the Arian faith he shared with Newton. Arians
denied that Jesus Christ was divine, and so they opposed
the prevailing orthodoxy of Trinitarianism, which links
together God, Christ and the Holy Spirit. Newton sensibly
kept quiet about being an Arian heretic, but Whiston was
exiled to London, where he earned his living as a lecturer,
writer and religious enthusiast. Although he was – and

still often is – mocked for his increasingly eccentric behaviour, Whiston was a pioneer who did much to introduce Newton's ideas into the English educational system. Even university professors found it impossible to understand Newton, who had sneered at 'little Smatterers in Mathematicks' and deliberately made his work accessible only to 'able Mathematicians'.[12]

Whiston first became interested in magnetism because he saw it as a way of making money. After he devised ingenious but totally impracticable methods for measuring longitude, he solicited money from wealthy backers to subsidise his search for a magnetic solution to the problem. Whiston also hoped that Halley's internal rotating loadstone would lend support to his own theory of the earth's origins. He had originally shot to fame with his book (diplomatically dedicated to Newton) suggesting that the earth had been hit by two comets. When it was first published in 1696, natural philosophers praised Whiston's *New Theory of the Earth* because it reconciled scientific and biblical accounts of the creation – for instance, although it included mathematical calculations, it also accounted for the Fall from Paradise and Noah's Flood.

A quarter of a century later, Whiston's status was declining, and he seized on Halley's ideas to vindicate his own. Now he declared that the first of his comets could well have been made of loadstone: its impact was, he argued, powerful enough to set the earth in motion, but could only transmit a slower rotation to the inner magnetic core. Whiston also combined Halley's theories with his own measurements of the earth's magnetic characteristics, although his aim was not to vindicate

Halley but to prove that the age of the earth is the same as that given in the Bible.

Instead of recording the patterns of magnetic variation, Whiston decided to focus on dip (now called inclination). Imagine standing a compass on its side. Its needle is now free to rotate in a vertical plane, but will sit at an angle to the horizontal – the angle of dip. At the Equator, the angle of dip is zero, and (in theory at least) at the north and south poles it is 90°. Equipping himself with longer and longer magnetic needles, Whiston travelled round England and France compiling maps to show how the angle of dip varies. He claimed to be able to measure longitude on land to within four miles, although critics laughed at the practical problems involved in transporting needles several feet in length. Using them at sea was even more complicated because they had to be lined up with the magnetic N–S meridian. One intrepid navigator complained that he had tried for several hours to measure the angle of dip in northern Canada, but every time he aligned his needle, his ice floe swivelled round. Somehow (bribery?) Whiston managed to persuade four captains to drill wide holes in the decks of their ships to accommodate his apparatus; hardly surprisingly, his suggestion of carrying out dip experiments in one of Halley's diving bells seems never to have been followed up.

If you flick through Whiston's little book on magnetism, you immediately notice that it's packed with arithmetical calculations. On closer inspection, you realise that he has not only used these lengthy sums to conclude that the diameter of Halley's internal sphere must be 1,150 miles, but that he has also performed a more unexpected feat: taking seemingly logical steps,

Whiston has used his magnetic measurements to estimate the date of Noah's Flood. He also worked out the age of the earth, but here he was helped by knowing the answer in advance. Biblical experts had already deduced from their close analysis of the scriptures that God created the world on 23 October 4004 BC.

According to Whiston, his own results meant that Halley's estimate of a 700-year magnetic cycle for the inner sphere was far too short – it should be 1,920 years. Blithely ignoring the immense practical problems of measuring dip at sea, Whiston insisted that his methods were far better than Halley's, and with great panache he provided a table showing how the dip would vary from 1720 up to 2040, when the earth's magnetic poles would return back to their current locations.[13] However, Whiston's book had little impact because forward-looking experimenters were far more interested in improving navigation than in mapping southern England or corroborating the Bible.

As well as attracting longitude opportunists such as Whiston, Halley's magnetic theory also appealed to antiquarian historians trying to date their ancient discoveries. While Whiston was touring the country with his long magnetic needles, Newton's friend William Stukeley was searching for evidence of England's vanished past. He was particularly fascinated by the stone circles at Avebury and Stonehenge, which he wanted to show were built by the Druids for their rites at the summer solstice. But first he had to explain why the entrances did not point directly towards the position of the rising sun. Stukeley found an ingenious if controversial answer – the Druids had used magnetic compasses to orient their buildings; however, because the variation had then been different, their needles had not pointed in the same direction as they would have done in the early 18th century.

Stukeley discussed this problem with Halley, who helpfully provided historical data and told him about his 700-year cycle. Stukeley also drew on the expertise of another elderly friend – Isaac Newton, who had carried out extensive research into the chronology of ancient civilisations. Leaning on these and other prestigious authorities, Stukeley claimed – very much to his satisfaction – to have proved magnetically that Stonehenge was built in 460 BC and Avebury in 1860 BC.[14]

Attaching precise numbers to Halley's magnetic model was tricky, but did attract Servington Savery, a reclusive inventor from Devon. Confirming details of Savery's personal life is difficult because, as an old handwritten family tree preserved in the British Library reveals, there was at least one Servington in every generation of the Savery family. However, it is clear that this particular one became interested in magnetism at school and was wealthy enough to continue experimenting for the rest of his life.[15] Rich, meticulous and perhaps educated at Oxford (or was that one of his cousins?), Savery was a gifted researcher, but he suffered from a major disadvantage – he lived a long way from London.

Unlike Halley, an expert networker who exploited his contacts, Savery operated at a distance. If he had been at the Royal Society in person to demonstrate his ideas, the long, detailed account of magnetism that he submitted in 1730 could have made him famous. Instead, it was read out sporadically and never even completed before the Fellows adjourned for the summer. Nevertheless, because

Savery's paper was published, it became an invaluable source of information for subsequent investigators.[16]

Savery relied on the power of arithmetic. Although magnetic measurements were rarely made more precisely than to a tenth of a degree, Savery's sums were calculated to seventeen decimal places. Accepting Whiston's figure of 1,920 years for a complete magnetic cycle, Savery constructed an ingenious device for rotating a six-inch magnetic terrella that served as a scale model of Halley's internal magnetic core. By counting this small sphere's revolutions with a twisted thread and timing it with an oscillating pendulum, Savery was able to work out the velocity of the actual inner globe.[17] To measure the diameter of the earth's internal magnet, Savery constructed another clever terrella machine designed to be used in conjunction with geometrical arguments.[18]

Whiston, Stukeley and Savery thought carefully but their criticisms had little impact. The vast majority of natural philosophers were happy to accept Halley's basic model because it explained the earth's magnetic characteristics far more satisfactorily than any other explanation. His critics used biblical arguments to maintain that at the centre of the earth lies a huge cavity that had been filled with liquid on the third day of the Creation. One of them sniped that Halley had invented 'an Hypothesis which can never be proved'.[19] But because nobody had been down there to look, Halley's suggestion was also hard to disprove – and it survived for well over a century.

* * * * *

Although Halley kept quiet about his elaborate subterranean speculations, he never forgot them. Over twenty years later, long after he had been made an Oxford professor, he was indoors 'at a Friend's House' and so at first was oblivious to the exciting events taking place outside – a spectacular aurora borealis was raging over southern England. Elderly gentlemen leapt out of bed, still wearing their nightclothes as they joined the awestruck crowds thronging the streets. According to the newspapers, many people were terrified, convinced that the end of the world was nigh, because up in the heavens 'they saw Armies engag'd, Giants with Flaming Swords, Fiery Comets, Dragoons, and the like dreadful Figures'. Two popular young Jacobite rebels had recently been executed on Tower Hill, and Halley knew that the aurora's red glow was traditionally described as showers of dripping blood.[20]

Once alerted, Halley stayed up until dawn observing the lights as they flashed eerily across the skies, constantly flickering and changing colour. Like other Enlightenment philosophers, he was determined to find a rational explanation and quell superstitious predictions of disaster. The Royal Society asked him to collate accounts from all over the country, and he described the whole sequence even though he had not seen the beginning himself. He always regretted those missed couple of hours, since it was over 80 years since this striking natural phenomenon had occurred so far south.

Halley's diagram (Figure 6) was also based on other people's reports as well as his own. He designed it not to give a snap-shot impression of the aurora at any one instant, but instead to present different features as they appeared in the sky at various times and places. By drawing in a stylised way, and scattering his diagram with

Figure 6: Edmond Halley's diagram of the aurora borealis of 6 March 1716. *Philosophical Transactions* 29, 1716, facing p. 389. (Cambridge University Library)

reference letters, Halley imposed geometrical order on to this nebulous, ephemeral phenomenon.

Halley would, he declared, give his readers 'a good History of the *Facts*' (his emphasis). He drew their attention to the small area on the lower right of his picture. Dismissing with a sneer 'the Conception of those that fancy'd Battles fought in the Air', he explained that this impression was due to the irregular motion of several small white columns he marked with Fs. Halley also carefully described how bright triangular cones (such as L and H) kept reappearing and blazing out in streaks of '*exotick Light*'.[21]

As well as assembling data from all over the country, Halley went back in time. Because he wanted to draw up laws of nature which used the past to predict how the world would behave in the future, Halley often engaged in retrospective research. For the aurora, he compiled information about all the similar spectacles that had been seen over England since the reign of Elizabeth I, well over a century earlier. His first conclusion was sensible but not terribly helpful – 'the Air, or Earth, or both, are sometimes, though but seldom and with great Intervals, disposed to produce this Phænomenon'. But why should this happen? And what were those coloured streaks and flashes? After rejecting a couple of obvious possibilities, Halley triumphantly produced a solution. These mysterious lights are, he announced, produced by '*Magnetical Effluvia*'.[22]

Effluvia were streams of particles that conveniently explained the inexplicable. Like Halley's mysterious aether, they were invisible and detectable only through their effects. Along with many other natural philosophers of the time, Halley suggested that some sort of 'subtle matter' exists – minute weightless particles which collectively behave like fluids yet can pour through solids with ease. As he himself put it, their 'Atoms freely permeate the Pores of the most solid Bodies, meeting with no Obstacle from the Interposition of Glass or Marble, or even Gold itself'.[23] Blessed with such elusive properties, effluvia defied detection, which meant that their existence could be neither conclusively proved nor disproved.

When faced with strange phenomena such as magnetism, electricity and light, natural philosophers struggled to find physical explanations that would sound different from the occult forces summoned up by magicians. Instead of imagining hidden powers somehow stretching through the universe, they preferred theories based on the mechanical interaction between particles. Although mechanical philosophers were forced to credit their subtle fluids with some weird properties, they still felt that effluvia – tiny invisible corpuscles – provided a more rational model of the cosmos than disembodied forces.

Halley drew analogies between the large-scale visible world of experience, and the concealed universe of minuscule atoms. He provided an illustration (Figure 7) showing how he had placed a magnetic terrella on a level surface and scattered fine iron filings around it. After some gentle tapping, the filings settled into patterns which he faithfully copied into this diagram. Victorian scientists would later use similar lines to argue for the existence of magnetic fields, but Halley drew different conclusions from his observations.

On Halley's interpretation, this experiment reveals the behaviour of subtle matter near a spherical loadstone as it

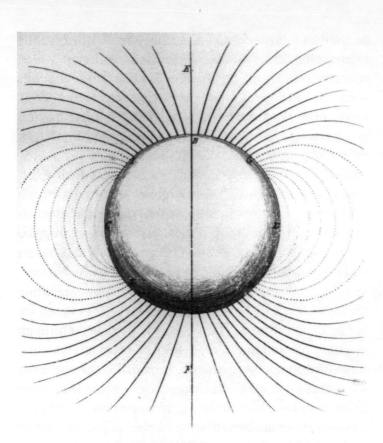

Figure 7: Edmond Halley's diagram of circulating effluvia.
Philosophical Transactions 29, 1716, Plate 2 facing p. 389.
(Whipple Library, Cambridge)

sweeps the iron powder into orderly shapes. Halley's next
step was to extrapolate upwards in scale from his terrella,
arguing that a magnetic effluvial fluid circulates around
the earth and penetrates it especially strongly near the
poles. With the help of that handy standby 'unknown
causes', Halley then proceeded to maintain that the
magnetic fluid rotating inside the earth produces a small

amount of light. And so he managed to account for the bright cones that appeared so splendidly on 6 March: they were, he said, spurts of magnetic fluid escaping through vents in the earth's crust. As Halley pointed out, this fitted nicely with his tentative suggestion – made twenty years earlier – that a special luminous medium fills the subterranean magnetic cavity.

Halley evaded criticism of this logical legerdemain by concluding modestly: 'I desire to lay no more stress on this conceit than it will bear.'[24] He never mentioned 'this conceit' again, but other authors had their own agendas which he could not control: they developed Halley's tentative conjectures in different directions.

* * * * *

Some experts cited Halley when they searched for magnetic explanations of the spectacular auroras that continued to appear unusually far south, although more often they pointed to electricity as the cause. But many natural philosophers were far more intrigued by extra-terrestrial life than by rare astronomical phenomena, and some of them picked up on Halley's suggestion of creatures existing down below the earth's surface – not nearly so strange an idea then as it might seem now.

Halley was exceptional not for believing that God had created other forms of life, but for putting forward this view so early, and also for focusing underground rather than in the heavens. Like Halley, many Enlightenment natural philosophers believed in the plurality of worlds – the idea that life exists on other planets and even on other solar systems so far undetected. By life, they meant intelligent beings – including angelic ones – rather than the microscopic creatures for which modern space scientists are searching.

To justify his arguments, Halley was creative with the facts. He claimed that 'it is now taken for granted that the Earth is one of the Planets, and they all are with reason supposed Habitable, though we are not able to define by what sort of Animals'.[25] This was an exaggeration. At that time, the reality of extraterrestrial life was by no means 'taken for granted' in England – even the existence of other planetary systems surrounding fixed stars was not accepted as fact. The first Newtonian to speak publicly about life elsewhere in the universe was Bentley, Halley's inquisitor. More than a year after Halley's lecture at the Royal Society, Bentley declared in a sermon (later published) that God had made innumerable stars 'for the sake of Intelligent Minds'.[26]

Although Newton remained equivocal, 30 years later life on other planets had become a standard part of Newtonian orthodoxy. The most important argument was provided by natural theologians. A truly magnificent God, they insisted, must have created many many universes supporting many many types of life. In 1721 Cotton Mather, an American preacher at the Royal Society, exclaimed '*Great* GOD, what a Variety of *Worlds* hast thou created! ... Who can tell what *Angelical Inhabitants* may there see and sing the *Praises* of the Lord! Who can tell what *Uses* those *marvellous Globes* may be designed for! Of these *unknown Worlds* I know thus much, *'Tis our Great GOD that has made them all.*' To provide an example of this munificence, Mather told his readers that 'Mr *Halley* allows there may be Inhabitants of the lower Story, and many ways of producing *Light* for them.'[27] Spreading through poems and sermons as well as scientific textbooks, the idea of extraterrestrial life had become firmly established by the end of the 18th century.

Halley coined an infernal pun to acknowledge that the centre of the earth was traditionally occupied by the fiery furnaces of hell. Providing referenced quotations for his well-educated readers, he justified his own suggestion about subterranean life by leaning on his classical predecessors: 'I am sure the Poets *Virgil* and *Claudian* have gone before me in this Thought, inlightning their *Elysian Fields* with Sun and Stars proper to those infernal, or rather internal, Regions.'[28] Unlike Catholics, Anglicans no longer believed that sinners were consigned to eternal torment, although preachers did continue to issue warnings about hell as a deterrent to too much enjoyment. But where was hell? This was a problem that troubled a few natural philosophers in the 18th century, including Whiston. He was among those who suggested that comets might act as mini-hells. But that posed another worrying problem: how could souls endure the torturously high temperatures when their comet passed near the sun?

Halley's model provided an ideal solution for Whiston, since its illuminated internal cavity provided a tolerably comfortable place where sinners would have plenty of time to repent before the Last Judgement. Whiston's blending together of theology and magnetism underlines that there were only blurred boundaries between religious, scientific, political and fictional pieces of writing during the 18th century. The most famous example is *Gulliver's Travels* by Jonathan Swift, who incorporated the latest philosophical controversies into his hero's fantastic tales. He created Laputa, a flying magnetic island; although Laputa was imaginary, its dimensions accurately mirrored those of the Royal Society's large terrella, and its hollowed-out interior recalled Halley's internal cavities.

Like Swift, other satirists also drew on the ideas of Halley and other natural philosophers. By imagining ideal civilisations, they could criticise their own. Some of them envisaged their heroes flying off to adventures on the moon or other planets, and – over a hundred years before Jules Verne – a few of them also ventured deep inside the earth. In one of these utopian fantasies, the narrator is magnetically pulled downwards to meet the inhabitants of Halley's central sphere. After a helpful philosopher has given him some special demagnetising ointment to unstick his feet, he walks around this interior world, which is lit by enormous jewels coating the inner surface of the earth's crust. Our traveller relates several uplifting lectures given by his hosts who are eager to point out the ignorance and insignificance of human beings, and stress how lucky he is to have learnt what lies at the centre of his world. The Jesuits must be very foolish, remarks his guide sarcastically, to insist that hell lies inside the earth – how would there be enough room for all those heretical Anglicans?

Another fantasy subterranean traveller, Niels Klim, was particularly important. The book's author, Ludvig Holberg, was Denmark's equivalent of Voltaire, a satirical dramatist who was well educated in theology and politics. Holberg had studied in Oxford for a couple of years while Halley was the Savilian professor of astronomy, and it seems clear that he knew about Halley's magnetic model of the earth's interior. Klim's adventures are still standard reading for Scandinavian school children; translated into English, Klim's underground experiences influenced famous writers such as Edgar Allan Poe, Giacomo Casanova and Mary Shelley.

After tumbling down a large cave in Norway, Klim discovers a subterranean universe illuminated – like Halley's – by a strange celestial aether. There he encounters bizarre peoples with alien social codes, and his farcical adventures resemble those of Lemuel Gulliver: although they are superficially humorous, at a deeper level they savage corruption and ridicule social foibles. Echoing Halley, Klim sneers at the subterranean residents who interpret his arrival as an extraordinary celestial phenomenon that 'prognosticated some impending Misfortune, a Plague, a Famine, or some other such extraordinary Catastrophe'. Klim eventually becomes rich and famous by introducing wigs to the inhabitants of Martinia (a thinly disguised France), but he first lands in Potu (Utopia backwards), a land inhabited by rational, mobile trees where – unlike in Europe – virtue is more valued than wealth, public appointments are made on the basis of ability, and sexual discrimination favours women rather than men.[29]

Mary Shelley recorded in her diary that one 'Teusday' in January 1817, she read Holberg's *Niels Klim*. She was right in the middle of producing her own famous novel, *Frankenstein*, and there are some striking similarities between the fantasy voyages of these two heroes. For instance, they both focus on the rejection of ugly out-siders, and they are both constructed as three concentric shells enabling different narrators to provide their own versions of events. Klim declares that he is driven by insatiable curiosity to explore the uncharted mysteries hidden underground; using similar language, Shelley describes how Robert Walton, an intrepid explorer, aims to satisfy his own fascination with unknown territories by visiting the north pole.

Frankenstein is often said to be the first work of science fiction, a new literary category that emerged as the boundaries separating factual from imaginary writing became more clearly defined. Yet Shelley herself had much in common with Holberg and the other utopian travel writers who preceded her. By reading about the fictional Niels Klim, she tapped in to Halley's suggestion of underground life, a supposedly scientific proposal presented to the Royal Society, but one with no more factual justification than Holberg's.

Halley also influenced Shelley more directly because his magnetic theories still prevailed in the early 19th century. Walton's main goal is, he writes to his sister in the opening pages of *Frankenstein*, to find 'the wondrous power which attracts the needle ... the secret of the magnet'. Shelley knew that at the same time as she was writing her fictional adventure story, the British government was being urged to send out very real Arctic explorers to investigate the earth's magnetic characteristics. 'Who could imagine', asked one admiralty official rhetorically, 'that the power of the magnet ... would lead to the discovery of a new world? and who can tell what further advantages mankind may derive from the magnetical influence, so very remarkable, so little understood?'[30]

Shelley's imaginary Walton spent his childhood devouring his uncle's library of early exploration books. These might have included Halley's account of his own voyages across the Atlantic – but the story of those exploits belongs to the next chapter.

· CHAPTER 4 ·
MAPPING MAGNETISM

So Geographers in Afric-maps,
With Savage-Pictures fill their Gaps;
And o'er unhabitable Downs
Place Elephants for want of Towns.

Jonathan Swift, 'On Poetry', 1733

As the 17th century drew to a close, the scientific gossip circuits buzzed with exciting news. 'Mr Hally has gott a ship from the government,' reported one astronomer, 'in which he has sett sail to goe round the globe on new discoverys, and the rectifying of geography.'[31] Halley had been concocting this project since his second magnetic paper was published, although almost six years went by before he finally left England on 24 October 1698. His was no ordinary ship – it was an Admiralty man o' war that Halley planned to take right round the world. Although he only ever managed to cross the Atlantic, he did make two pioneering voyages of state-funded exploration.

By the time that Halley's *Paramore* eventually set sail, it carried eight guns, over twenty men, and Admiralty instructions largely composed by Halley. He told himself 'in all the Course of your Voyage, you must be carefull to omit no opportunity of Noteing the variation of the Compasse, of which you are to keep a Register in your Iournall'.[32] Despite appearances, this was not a purely scientific expedition – Halley was also ordered to map as many Atlantic islands as possible. Just as when James

Cook was later sent to colonise the Pacific under the cover of making astronomical measurements in Tahiti, Halley was meant to be searching out good trading and military opportunities for his country. When he encountered the small island of Trinidada, he not only surveyed it – he also took possession.

During his long journey westwards, Halley recorded many scientific observations (as well as stopping for a few days to pick up some wine at Madeira), but he was hampered by his ship's behaviour. It was slowed down by the barnacles which kept accumulating on the bottom, and was also unstable, making it difficult for Halley to read his instruments. As his lieutenant, Edward Harrison, complained: 'by reason of ye smallness of ye ship it rowld and moved so fast y^t he could not observe so exactly as they use aboard ye bigger and steadier East India Merchant men.'[33]

In addition, Harrison was himself a major problem. An experienced naval officer, he resented serving under a landlubber like Halley, and also bore a personal grudge – a few years previously, Halley had strongly criticised Harrison's own book suggesting ways of finding longitude magnetically. Within a few weeks, the crew members were listening to Harrison rather than Halley, and after five months the *Paramore* turned round and headed back for England. The atmosphere on board was horribly tense, and Halley complained bitterly to the Admiralty. However, he seems to have kept quiet in public: perhaps there was some foundation in Harrison's scathing accusations of incompetence.

A few months later, minus an Admiralty lieutenant but helped by extra seamen and a one-armed boatswain,

Halley set off again. This time he was away for a year, from 1699 to 1700, and often boasted that, apart from a small boy who fell overboard, he brought back all his men alive – no mean feat at this time, when pirates, storms and sickness made oceanic voyages very dangerous. Halley also took many measurements of longitude and magnetic variation, using his own astronomical instruments and azimuth compasses. Harrison would surely have approved, since he had denounced the naval ignorance of his 'Honourable *Company Masters*: little do they know, how many Ships have been lost for want of a better knowledge of the *Variation*.' Harrison derided the Navy for supplying their ships with compasses that were 'Old and Rusty, and good for nothing, except to throw overboard'.[34]

In Halley's own chart (Figure 8), the dotted line shows the route of his second trip across the Atlantic, setting out towards Rio de Janeiro, following a figure of eight loop via St Helena up to Bermuda, and finally heading almost due east back to England. As soon as he got back, Halley embarked on another campaign of self-promotion.

* * * * *

After only a couple of months, Halley produced his first Atlantic chart (Figure 8), which the Fellows of the Royal Society proudly hung on the wall of their meeting room. The original has now vanished, perhaps destroyed by tearing under its own weight or from years of accumulated smoke and soot, but – like other maps – it was a declaration of possession. Transferring portions of the globe to a piece of paper was the next best thing to owning them. By scrutinising Halley's chart, the armchair

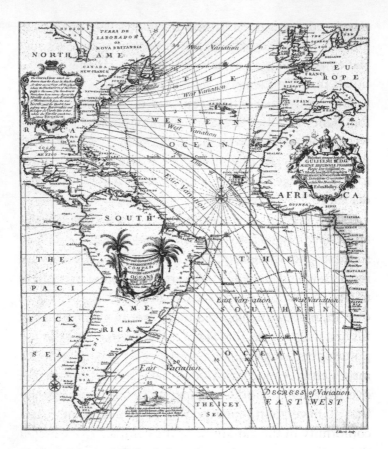

Figure 8: Edmond Halley's first chart of magnetic variation, 1701. (British Library)

voyagers of the Royal Society could vicariously share his experiences.

This was a map with messages. Its diplomatic dedication to King William lies inside the African cartouche, which is adorned with three female muses holding up a compass and other navigational instruments (to be used, of course, by men). The family lolling under the misplaced palm-trees in South America conformed to English prejudices about indigenous peoples. Because Halley wanted to advertise his exotic experiences, he noted that the *Paramore* sailed far south into 'The Icey Sea'. Once there, when he could see through the freezing fog, Halley negotiated gigantic icebergs, collected strange weeds and marvelled at penguins, whales and 'Diving Birds with Necks like Swans' – presumably the source for the strange creatures shown swimming near Tierra del Fuego.[35]

Halley needed to make sure that Admiralty officials appreciated how fully he had justified their financial support, and so he sent back letters promising that his research would 'appear soe much for the public benefit as to give their Lord[sps] intire satisfaction'. He also obeyed his own injunction, delivered almost twenty years earlier in his first magnetic lecture at the Royal Society, that natural philosophers should search for 'some general rule' to make sense of magnetic phenomena. While he was still overseas, Halley confidently maintained to the Admiralty that 'I have found noe reason to doubt of an exact conformity in the variations of the compass to a generall theory.'[36]

Being creative with the truth is a modern expression, but Halley did make sure that his boast would appear justified. He was so determined to find an orderly system underlying the earth's magnetic characteristics that, on the basis of only 150 readings, he drew suspiciously smooth curves across the entire Atlantic joining together points with the same magnetic variation. As he explains inside his North American cartouche, the double line marks zero variation, where the compass needle pointed directly towards magnetic north. Towards Europe, the variation was westerly, and towards the Americas, it was easterly.

Halley was the first person to introduce what are now called 'isogonics', curves that link together abstract, measured quantities such as temperature or pressure. In contrast with contour lines, isotherms and isobars are not direct representations of the earth's features. Although such isogonics are now familiar, it was only in the 19th century that these 'Halleyan lines', as they were then known, became standard. They enabled Halley to impose a neat pattern onto the magnetic observations that had previously appeared so chaotic. He was systematising the world with the same approach as the Enlightenment encyclopaedia compilers, who often used mapping metaphors to describe how they were organising facts into territories and domains.

Within a year, Halley had produced yet another innovation – a magnetic chart not just of the Atlantic, but of the whole world. He designed it to look modern and authoritative. The appealing but old-fashioned pictorial cartouches had been eliminated, and a specially composed Latin poem celebrated the commercial benefits of the compass which 'brought mutual products to remotest lands'. Aiming to entice mariners, Halley added printed instructions about reading the chart at sea. He claimed that navigators could use it either for estimating longitude (by comparing their variation with that shown on the chart) or for correcting compass readings (by looking up the variation at a known longitude). He also appealed to natural philosophers by showing neat patterns of regular Halleyan lines.

Several versions of this second chart were published. Yet despite their impressive appearance, Halley's charts suffered from some serious shortcomings. For one thing, since he had not himself sailed round the world, Halley had to use measurements recorded by other people. But he had no guarantee of their accuracy – and in any case those made years earlier would no longer be correct since the variation was constantly changing. In addition, because coasts were far better mapped than interiors, he had very few readings on land. To overcome these difficulties, Halley decided that three points were enough to justify a smooth curve. Sometimes he even made up suitable results to fill a gap – 'I never observed myself in those Parts; and 'tis from the Accounts of others, and the Analogy of the whole, that in such cases I was forc'd to supply what was wanting.'[37] Even when his charts were updated in the middle of the 18th century, these problems still remained.

Maritime men were scathing. They complained that Halley's charts were too small, became rapidly out of date as the variation changed, and were inherently of least value along the most important trade routes in the northern Atlantic, since there the lines of equal variation run almost due east–west and so give little hint of longitude. One London navigator sneered that the charts were 'laid down from very uncertain observations, and the rest from the mere imagination of the draughtsmen … but experience quickly informs us they are only pictures, the creatures of the brain'.[38]

But the Fellows of the Royal Society continued to congratulate themselves on Halley's pioneering achievement. In order to promote itself, the Society needed to demonstrate its contributions to British trade and empire. Britain was pursuing an aggressive policy of commercial expansion in the middle of the 18th century, and the Royal Society's President stressed that magnetic research would benefit the economy by making it easier to import foreign luxuries. The Fellows were, he claimed, disinterested contributors to the public welfare 'earnestly wishing to become day by day, more and more remarkable for their constant application to promote useful knowledge ... to the advancement of Science, and to the general benefit of the Publick'.[39]

Defoe pinpointed the main reason why Enlightenment philosophers were interested in magnetism: money. 'By the discovery of the Magnet and the use of the Compass', he boasted, 'Men were particularly qualify'd to visit remote Countries, and make both Discoveries and Improvements also in Trade and Plantation.'[40] Science may have meant progress – but it also meant profit.

* * * * *

Harrison was aggrieved because Halley had taken over his job, but this transfer of knowledge and status from experienced practical men to the new scientific investigators was not unusual. During the 18th century, Enlightenment natural philosophers continued to convert navigators' traditional skills into a new science of magnetism. Seasoned mariners protested at this invasion of their territory. They stressed the value of experience over newfangled techniques and instruments. The Liverpool Dock Master, for instance, complained that relying on calculations could be positively dangerous in an unpredictable environment governed by fluctuating tides, winds, and magnetic variation.

Nevertheless, magnetic researchers set themselves up as specialised experts, socially and intellectually superior to the experienced seafarers whose knowledge they had acquired. By the beginning of the 19th century, land-based scientific lecturers were giving talks and writing books about magnetism. They aimed to earn not only their living but also a higher social status in recognition of their control over the natural world. Figure 9 shows the magnetic diagrams compiled by Thomas Young, a brilliant polymath at the Royal Institution who revolutionised optics and also helped to decode the Rosetta stone. Because paper and printing were still expensive, his illustrations are crammed together as an economy measure, and the three in the bottom right-hand corner are on another topic. Young's five globes advertise how knowledge about the earth's magnetic patterns had expanded since Halley's time. The upper three compare the recorded variations in 1700, 1744 and 1790, while the lower two are more hypothetical; based on modelling experiments, they show the patterns of dip that would be expected if different magnets were placed at the earth's centre.

Natural philosophers had much to learn from maritime men, the true magnetic experts. Throughout the Enlightenment, ships' captains compiled magnetic measurements and sent them in to the Royal Society. They had mixed motives. Many of them wanted to make some money, so they marketed updated versions of Halley's chart, or took advantage of official backing to

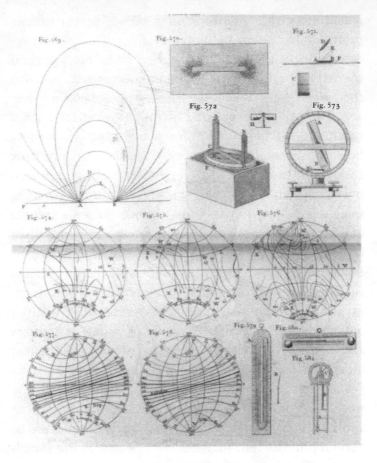

Figure 9: Magnetic diagrams. Thomas Young, *A Course of Lectures on Natural Philosophy and the Mechanical Arts*, London, 1807, vol. 1, plate 41. (Whipple Library, Cambridge)

advertise compasses they had invented; others seem to have been gratified simply by seeing their name in print in the *Philosophical Transactions*. It was in the interests of the Admiralty and private shipping companies to sponsor magnetic observations, since more detailed knowledge would hopefully result in safer journeys and greater profit. When the famous navigator George Anson captured Spanish ships on his celebrated voyage around the world, he seized not only his victims' gold but also their secret logbooks packed with magnetic information.

Young's sketchy maps also reveal another important technique that men of science derived from mariners – how to summarise information visually rather than listing it in long pages of figures. Unlike natural philosophers, navigators were used to reading charts and carrying out three-dimensional trigonometrical calculations. Magnetism is not the only example of how visual languages entered the sciences from practical occupations. Geology was also a new subject at this time, and geologists learnt how to represent their results from experienced surveyors and mine workers. Similarly, botanists and meteorologists were preceded by farmers and artists who had discovered how to distinguish different types of trees and clouds.

Magnetic maps transported data from one side of the world to another, and also from one community to another. Navigators pored over them in their cabins, natural philosophers perused them in their studies, map publishers debated the profitability of updating them, and students struggled to understand them. Along the way, Halley's original chart was reinterpreted. Magnetic experts carried out thought experiments, suggesting that the earth's nucleus curved in an S-shape to reflect Halley's lines snaking across his chart, or (as in Young's lower two diagrams) envisaging how different internal configurations of imaginary magnets would generate particular patterns on the surface. These speculations had a lasting influence. When Michael Faraday fused together electricity and magnetism in the early 19th century, he adapted Halley's curved isogonics to represent lines of force in the earth's electromagnetic field.

Young showed modernised equipment for measuring variation and dip (572 and 573 in Figure 9). Although magnetic instruments continued to be designed on the same principles as those used by Halley, Whiston and their contemporaries, by the time that Shelley described Walton searching for the earth's magnetic secrets, they had become far more accurate. She knew that campaigners at the Royal Society were persuading the government to fund expeditions using such high-quality precision instruments. They wanted to update Halley's chart by taking magnetic measurements all over the world; once the Napoleonic wars were over, explorers embarked on magnetic crusades to chart the entire globe.

This was, of course, a collective endeavour, but some changes did take place through the initiative of particular individuals. In Enlightenment England, the country's most eminent magnetic expert was an eccentric doctor called Gowin Knight. It was because of his desire to become rich and famous that traditional magnetic instruments started to change – and so the next section of this book focuses on Knight, an unfamiliar but crucial hero in the story of magnetism.

PART TWO
KNIGHT'S NAVIGATIONAL NOVELTIES

Yet powerful it urges: thirst of gold,
And lust of sway, and fiercer than these fiends,
Relentless superstition; not alone
The love of knowledge, and the towering step
Of Virtue, all divine, have led the way.

Capel Lofft, *Eudosia; or, a Poem on the Universe*, 1781

Gowin Knight (1713–72) is not a famous man. Even most specialists in the 18th century have never heard of him. He was also apparently extremely unpleasant. Only fleeting references to his personality survive, but one of the most telling is by the poet Thomas Gray, who commented on the sour atmosphere among the staff who worked under Knight at the British Museum. The curators had, he reported, 'broke off all intercourse with one another, & only lower a silent defiance as they pass by'. Hardly surprising that they were so resentful – Knight had 'wall'd up the passage to the little-House [toilet], because some of the rest were obliged to pass by one of his windows in the way to it'.[1]

Interesting people are not necessarily nice to know. Knight has been pushed into obscurity, but there are two major reasons why it is worth rescuing him. Most obviously, he was well known at the time. Our perception of the past is often overshadowed by relatively minor figures who have left large archives of manuscripts for researchers to explore. In contrast, Knight had many acquaintances but few friends, and all but a handful of formal letters have vanished. Yet although it is now impossible to do more than reconstruct the skeleton of his own life, he was an important player in Enlightenment England whose influence spread out through maritime as well as scholarly communities.

In addition, Knight deserves attention because he represents the countless 18th-century opportunists who founded our modern consumer society. The economy was booming, and for the first time ever, enterprising individuals could become powerful by creating fortunes rather than inheriting them. The best-known example is Josiah Wedgwood, who persuaded his customers to break the thrifty tradition of centuries. Instead of handing down their china through the generations, wealthy women should – according to his seductive advertising literature – give their plates away to the servants and buy the latest fashionable designs for themselves.

Knight was no Wedgwood, but he was intelligent, ambitious and avaricious. An adept social climber, Knight started low and finished high. He was a showman who turned every opportunity to his own advantage, but he was also an ingenious, dextrous man who invented some fine instruments. The secret of his success lay in his marketing techniques: he was a skilled salesman who knew how to promote himself as well as his scientific products.

Knight was not alone. All over the country, not just in London, thousands of experimenters were working out new ways to earn a living. Fired by hardship as well as enthusiasm, they taught, wrote books, invented instruments, staged public performances and lectures. Collectively, these entrepreneurs made science respectable. At the beginning of the 18th century, natural philosophers and inventors were figures of fun, mocked as bumbling virtuosi and impractical projectors. A hundred years later, no young man or woman could call themselves educated unless they knew some science.

Knight designed and sold magnets and compasses because he wanted to become rich and famous. (Or is that assessment perhaps too harsh? Could there also have been some truth in his claims that he had the welfare of the nation at heart?) He liked to talk about himself as England's greatest magnetic inventor – not an unreasonable boast. But looking back, another type of innovation seems to have been far more significant. Knight helped to introduce a social invention that ultimately had far more impact than his revolutionary compasses – the scientific career.

The word 'scientist' was not invented until long after Knight's death, in 1833, when a new word was coined to describe a new type of professional man. By the mid-19th century, clever young school-leavers (if they were male) could choose to embark on a scientific career by studying science at university, joining scientific societies and writing articles for the rapidly expanding number of academic journals; most importantly of all, they could expect to be paid for research and teaching.

Such openings were unavailable to Knight and his contemporaries. However, without their initiative, salaried jobs would not have existed for Faraday and the other great scientists of the Victorian era. Knight's story is fascinating because it reveals a vanished way of life that he himself helped to eradicate.

· CHAPTER 5 ·
WHO WAS GOWIN KNIGHT?

I suppose you are not so rigorous or cynical as to deny the value or usefulness of natural philosophy; or to have lived in this age of enquiry and experiment, without any attention to the wonders every day produced by the pokers of magnetism … I offer you the honour of introducing to the notice of the publick, an adept, who having long laboured for the benefit of mankind is not willing, like too many of his predecessors, to conceal his secrets in the grave … I hope no man will think the makers of artificial magnets celebrated or reverenced above their desserts.

Samuel Johnson, *The Rambler*, 1752

Benjamin Franklin and his fellow American experimenters were kept up-to-date with the latest London research through a network of diligent Quaker correspondents. In 1745, there was some exciting news to report. 'Hither to I have wrote only to blot paper,' gushed one enthusiastic letter writer, 'but now I tell you some thing new Docʳ night a Physition has found the Art of Giveing Such a magnetic power to Steel that the poor old Loadstone is putt quite out of Countenance.'[2]

Then 32, 'Docʳ night a Physition' was still unknown, but his rise was meteoric. Within a few years, Dr Gowin Knight was a Fellow of the Royal Society, had published a book, negotiated a contract with the Royal Navy, and been appointed the first director of the new British

Museum. His nearest approximation to a close friend was Benjamin Wilson, an electrical experimenter and portrait artist, who produced the only known picture of Knight (Figure 10). Dressed in sober clothes and wig, Knight is flanked by his two major achievements: his book on the

Figure 10: Gowin Knight, painted and etched by Benjamin Wilson, 1751. © Royal Society

table behind him, and – placed more significantly at his front right – his navigational compass.

* * * * *

Like many of his entrepreneurial colleagues, Knight was a clergyman's son. Apart from the existence of one sister, little is known about his family, but at some stage they moved from Lincolnshire to Leeds, where Knight went to the same school as Wilson and John Smeaton, the engineer celebrated for designing the Eddystone light-house. When Knight was eighteen, he won a scholarship to Magdalen Hall at Oxford: several of the wealthier colleges had halls to accommodate the swelling numbers of poor young men from medical and clerical families.

Oxford and Cambridge, England's only two universi-ties, were not renowned for their scholarship. Many of their wealthy aristocratic students were more interested in drinking, gambling and hunting than in serious study; lecturers were rarely in residence, and the examinations were often a formality. A foreign visitor was horrified at the mould growing on the library books at Magdalen, and the famous historian Edward Gibbon lamented that his spell at Oxford was 'the fourteen months the most idle and unprofitable of my whole life'.[3]

For four years, Knight survived by acting as servant to the richer students – an initiation rite that Isaac Newton had also endured at Cambridge. But life improved when Knight won a more lucrative scholarship promoting him to Magdalen College, where he stayed for a further six years, studying medicine and natural philosophy. Instead of spending all his time with his richer colleagues round card tables and in inns, Knight decided to start experimenting with magnets. Perhaps he was inspired by meeting an undergraduate at another college, Servington Savery, a son of the confusingly named Devonshire recluse whose long paper had been published in the *Philosophical Transactions* in 1730, the year before Knight went up to Oxford.

Savery (senior) was a magnetic expert, but he failed to make an impact because he operated from a distance. In contrast, Knight succeeded by adopting self-promotional tactics as soon as he left Oxford. His first step was to secure patronage, and he somehow persuaded the former President of the Royal Society, Sir Hans Sloane, to recommend him to his successor, Martin Folkes. Enticed to Knight's lodgings, Folkes was impressed to see Knight's artificial magnets lift heavy bunches of iron keys; he was still more astounded when Knight mysteriously managed to increase the strength of loadstones. Even at this early stage in his career, Knight was worried that someone might steal his discoveries. He would disappear into his study with small pieces of loadstone, and emerge theatrically a few minutes later to demonstrate that he had somehow made them stronger.

With Folkes' backing, Knight could only move upwards. He gave a guest talk at the Royal Society in 1744, the following year was elected a Fellow, and in 1747 was awarded the Copley Medal (the Society's most prestigious award) for his new compass. Knight worked hard to infiltrate London's scientific elite. Soon he was a member of the Royal Society's inner circle, enjoying sumptuous meals at their special dining club; by 1751 he had been elected to the Council and was running for the post of secretary. Surviving letters reveal some of the political machinations behind this important election: supported by Lord Northumberland, Knight won 76 votes, but he was defeated by the more powerful lobby organised by the Lord Chancellor.

Despite this high-flying life, money remained a big problem. Although – strangely – Knight never took his final qualifying examinations at Oxford, he did practise as a doctor. To attract wealthy patients, Knight moved to fashionable Lincoln's Inn Fields, and he lost no opportunity of advertising this prestigious address. So when he told the Fellows about his experiences during the London earthquake of 1750, he carefully included the gratuitous information that his neighbour was the Duke of Newcastle, one of England's most powerful politicians.

Knight's medical schemes do not seem to have been very successful. Written references to only one patient have survived – his sister, whom he treated for a mysterious fever. Since he admitted that she died, he presumably abandoned any hope of trying to earn his living in the competitive medical market. A change of address often indicates a fresh start, and Knight moved in nearly next door to the Royal Society, taking rooms in Crane Court, a small alleyway lying at the heart of Fleet Street's instrument-making business. He tried to boost his income by selling his own diverse but ingenious inventions, which included Venetian blinds and a navigational device for measuring the depth of oceans. But Knight found that it was far more profitable to convert his magnetic experiments into money-making projects.

Using the Royal Society as a platform for self-promotion, Knight set out to corner the international market in magnets and compasses. Taking advantage of the coveted initials FRS following his name, Knight published articles, sent free samples to the Royal Society of Paris, and arranged for his products to be sold all over Europe by George Adams, one of London's most famous instrument-makers. When his magnets were omitted from the preliminary design for a diploma displaying the Royal Society's achievements, Knight pestered the Council until a new engraving was made. He always targeted the top end of the market, setting high prices on instruments that were beautifully crafted.

Knight's Fellowship also gained him an entrée into Britain's most lucrative magnetic market – the Royal Navy. Conveniently for Knight, his patron Folkes was a Commissioner of Longitude, and the head of the Admiralty – George Anson – was a Fellow of the Royal Society. After Knight had treated Anson to a repeat performance of his magnetic virtuosity, he must have felt confident of winning a naval contract. With Anson's backing, he visited the dockyards several times and convinced the Navy to buy his magnets and compasses. Years later, Knight was still mining every opportunity to make a sale. It can hardly be coincidence that after a briefing discussion at the Royal Society, the Pacific explorer James Cook wrote to the Admiralty: 'Doctor Knight hath got an Azimuth Compass of an Improv'd con[s]truction which may prove to be of more general use than the old ones; please to move my Lords Commissioners of the Admiralty to order the Endeavour Bark under my command to be supplyed with it.'[4]

To tap the international market of natural phil-osophers, Knight wrote a Newtonian textbook on magnetism, investing his own money in a high-quality product. Since paper was expensive, large books automatically demanded respect. Yet despite his substantial outlay, there were no rave reviews. Even the title suggested that a long, turgid work lay between the handsome leather covers – *An Attempt to Demonstrate, that all the Phænomena in Nature may be Explained by Two Simple Active Principles, Attraction and Repulsion: Wherein the Attractions of Cohesion, Gravity, and Magnetism, are Shewn to be One and the Same; and the Phænomena of the Latter are more particularly Explained*. Franklin phrased his verdict with exquisite tact, regretting that he had never found the 'Leisure to Peruse his Writings with the Attention necessary to become Master of his Doctrine'. Another reader was more blunt: Knight 'calls Old Discoveries by New Names, and deduces Corollaries till he loses all Sight of his Proposition'.[5] Nevertheless, Knight's book was well known, and ran to a cheaper second edition. Just like now, even problematic publications helped to promote their authors.

Personal contacts were invaluable. Once when Wilson was sheltering from the rain in Adams' shop, he entertained a fellow customer with tales of Knight's magnetic exploits. In his self-vaunting autobiography, Wilson boasted how he held his listener's attention. 'Now by this introduction,' he continued, 'I not only obliged [him], but served my friend the Doctor', who ended up making £250.[6] Knight's colleagues reinforced his publicity campaign by recommending his products in their own books on magnetism. 'These Compasses', one Fellow wrote in a free advertising puff, 'are made by *George Adams*, mathematical Instrument-Maker to His Royal Highness the *Prince of Wales*, and before they pass out of his hands, are examined and attested by the said Doctor *Knight*, whose certificate is fixed up to the Cover of the Box; without which they are not to be depended on.'[7]

As Knight's status rose, he started extending patronage to junior men. Helping one of them, a schoolmaster called John Canton, turned out to be an unwise move because as his protégé became successful in his own right, the two men became locked into a bitter priority dispute that reverberated for the rest of the century. Knight had more luck with Smeaton, whom he rescued from a love affair that was going horribly wrong. Smeaton remained Knight's assistant for several years until his own career took off, when he became far more distinguished than his former protector.

* * * * *

Money remained a major difficulty until, after ten years, all the networking paid off: Knight became one of the earliest natural philosophers to acquire a paid job. Even the position itself was new – Principal Librarian of the British Museum, the institution founded after Sloane bequeathed his vast collection to the nation. The organising committee was stacked with Fellows of the Royal Society, who exchanged letters in code as they discussed the sensitive question of the Museum's directorship. After Knight's friends had arranged some helpful dinner invitations, his prestigious referees convinced King George III to choose Knight over two rivals for the post.

Knight embarked on this early form of scientific career in 1756, and he remained at the British Museum until sixteen years later (when he was found dead in his office).

He played an important – if somewhat obstructive – role in the first stages of what are now known as programmes for the public understanding of science. In exchange for his secure salary, Knight became a live-in caretaker responsible for arranging the exhibits and superintending the visitors. Sloane's legacy included many plants, animals and other scientific specimens, and Knight supervised their display. He also agreed that the gardens should be laid out in the controversial classification system recently introduced by the Swedish botanist Carl Linnaeus, and still in use today.

But although Knight was in a position of power, he found it hard to concentrate on promoting science rather than himself. When he conducted distinguished visitors round the building, he made sure that they took a detour into the room where he had installed his large magnetic machine (Figure 15, page 105). As he continued to market his magnets and compasses, he did enlarge the Museum's collections of objects and books, but he also clamped down on the opening hours. Knight alienated visitors by restricting access, he antagonised his staff by making unreasonable demands, and he infuriated his superiors by exceeding his budget.

Now occupying an influential position, Knight sat on committees, met statesmen and aristocrats, and was recognised when he attended public lectures. But earning a salary encouraged Knight to nurture ambitions beyond his means. Along with Wilson, he gambled in Cornish mining projects, but discovered that his £200 a year was not enough for him to repay the thousand guineas that he lost (a guinea was £1.05); his debt was covered by one of his investment partners (a respectable Quaker doctor who presumably wanted Knight to keep quiet), although another of these financial colleagues was later accused of embezzling the Royal Society's funds.

Becoming famous did not result in becoming popular. Knight was a familiar figure at meetings of the Royal Society and dinner parties, but he seems to have been invited out of duty rather than affection. Blocking up the corridor to a museum's lavatory is not the way to win friends and influence people. Knight had the reputation of being short-tempered, reclusive and mean. Those deficiencies might not have mattered quite so much if he had been more open about his scientific discoveries. Knight was notoriously secretive. What probably started out as an instinctive drive for self-protection against his competitors turned into an adamant refusal to share his discoveries, an attitude flagrantly contradicting the Royal Society's ethos that scientific research should benefit the world, not the individual. Wilson reported that Knight refused to divulge his magnetic techniques even when offered as many guineas as he could carry, and this covetous reputation spread well outside the Royal Society. Samuel Johnson was just one among many who sneered that Knight's lust for money had overcome his thirst for knowledge.

All over Europe, writers repeatedly criticised Knight's secrecy. This is because they were attacking not only Knight's personal character, but also the very notion of making a profit from scientific inventions. Along with many other experimenters who needed to earn money, Knight was trying to change what it meant to be a natural philosopher. Most natural philosophers had been born rich, and so it was easy for them to denounce entrepreneurs who sold their skills and accuse them of debasing science by exploiting God-given natural wonders.

But for poorer researchers, marketing their instruments and their expertise was the only way to continue working. To stake his claim to his profits, Knight filed patents for his inventions: the records show that he was the first person ever to patent a compass. Many people still believed that new devices were revealed by God rather than being created by human ingenuity, and the patent system was little used because it gave inventors only poor protection against plagiarism. By taking out patents, Knight established an important precedent: during the great Victorian age of invention, patenting became standard practice.

When Smeaton proposed to support himself by giving lectures on electricity, he defiantly defended himself in advance – 'I dont see wheres ye harm as there is no fraud or Dishonesty [in] it', he protested.[8] But many commentators disagreed. It was wrong, they insisted, for philosophical entrepreneurs to profit from science. Opposed to commercialisation, they argued that aiming to become rich and famous must inevitably bias results and tempt researchers away from the true path to knowledge. This point was put most eloquently in an extraordinary horticultural metaphor hurled at Knight by a writer who was an expert on gardens as well as instruments: 'The Plants and Trees of the Gardens, of the Arts and the Sciences, cultivated by the Dung of Ambition, and nourished with the Waters of Interest, are very subject to be blasted by the Whirls of Error, and sometimes stunted by the Weeds of Imposition.'[9]

Knight may have been fed too much 'Dung of

Ambition', but he was an important pioneer in the drive to establish the professional, paid science that flourished during the Victorian era. Nineteenth-century scientists who looked back at the Enlightenment never acknowledged this particular debt to Knight and his colleagues. Faraday and other experts did, however, recognise the lasting importance of his magnets and his compasses. Knight may not be one of science's great heroes, but in specialist circles he was – and still is – recognised as a major magnetic predecessor.

· CHAPTER 6 ·
MARKETING MAGNETS

Then thrice and thrice with steady eye he guides,
And o'er the adhesive train the magnet slides;
The obedient Steel with living instinct moves,
And veers for ever to the pole it loves.
'Hail, adamantine STEEL! magnetic Lord!
King of the prow, the plowshare, and the sword!'
Erasmus Darwin, *The Economy of Vegetation*, 1791

Magnets meant money. In 1756, the Countess of Westmorland wanted to make sure that her husband was appointed as the next chancellor of Oxford University, so to impress King George II with the couple's loyalty, she encased a giant loadstone in a copper coronet and donated it to Oxford's Ashmolean Museum (Figure 11). Her expensive – and very heavy – gift is still on display (and her man did get the job).

Exceptional loadstones were identified by their owner's name, large ones being famous for their size, and small ones for their strength. The King of Portugal's enormous lump of magnetic mineral weighed 30 pounds, whereas the tiny chip in Newton's signet ring was renowned for being able to support 250 times its own weight. Social climbers who needed unusual bribes for currying favour with their patrons might pay several hundred pounds at auctions for a choice loadstone. The recipient would add this natural curiosity to his (and sometimes her) cabinet

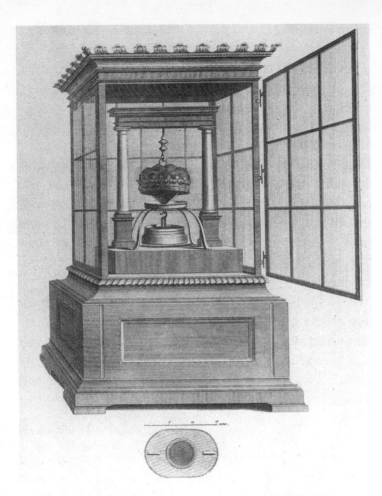

Figure 11: The Countess of Westmorland's loadstone.
© Museum of the History of Science, Oxford University

collection, and expect visitors to admire these strange objects, valued for their rarity. Compared with shells, fossils and stuffed animals, loadstones did at least do something – a travelling Englishwoman in Vienna admired the Emperor's 'small piece of loadstone that held up an anchor of steel too heavy for me to lift. This is what I thought most curious in the whole treasure.'[10]

Less spectacular models, similar to the Russian ones shown in Figure 12, were more typical (for drawings of other examples, look back to the playing card shown in Figure 1, page 4). Carved into rectangular blocks, these armoured loadstones are protected by decorative silver cases and strengthened by pieces of iron; they have been dipped into iron filings, which cluster round the iron caps. To armour a piece of magnetic ore, it was first shaped into a rough block and then carved flat along the sides of its poles; next it was mounted in two L-shaped iron blocks, with the polar faces vertical. This assembly was then covered with a metal casing – often silver, as here – and given a ring or handle to lift it. The ends of the iron blocks protrude down as two magnetic feet, which are protected by an iron bar or keeper when the loadstone is not being used.

Figure 12: Loadstones. (Science Museum, London)

Sometimes natural loadstones were shaped into spherical terrellas, and might even be marked with lines of latitude and longitude. The Fellows of the Royal Society were particularly proud of Christopher Wren's six-inch terrella, which had been set into a table-top and surrounded by 32 compass needles to simulate the earth's magnetism. The Society's loadstone even impressed Ned Ward, a satirical and often savage London journalist, who reported that 'it made a paper of Steel Filings prick up themselves one upon the back of another, that they stood pointing like the Bristles of a *Hedge-Hog*; and gave such Life and Merriment to a Parcel of Needles, that they danc'd the *Hay* by the Motion of the Stone, as if the Devil were in them.'[11]

Although natural loadstones fascinated tourists and aristocrats, navigators and experimenters found them awkward to work with. They were expensive, their quality varied, they were heavy and unreliable. Craftsmen had tried to make artificial substitutes from iron bars, but the soft iron rapidly lost its magnetic power. During the first half of the 18th century, metal workers found better ways of making steel. Although steel is hard to magnetise, it does retain its strength. Gowin Knight, that clever opportunist, took advantage of this new material to make and market a new product – artificial magnets.

* * * * *

Unlike electricity, magnetism was not a strange new force invented by natural philosophers, but had long been part of everyday life. The toy industry had not yet been invented, but children were given small pieces of load-stone to play with (much appreciated for breaking adults' watches). In addition, familiar objects mysteriously became magnetic – crosses on churches, window-frames of houses, workmen's tools, gentlemen's swords and keys. Through making artificial magnets widely available, Knight helped to make systematic, accurate research a realistic possibility. By the end of the 18th century, a new scientific discipline had been forged – magnetism.

Since Knight was so secretive, we have no detailed records of how he made his artificial magnets. It's clear that he developed several approaches, including some sort of chemical process in which he heated powdered loadstone to produce tiny but powerful magnets. That original recipe died with him, but the general principles of his technique for magnetising steel bars soon became public knowledge. Although the details varied, many experimenters used similar procedures.

Figure 13 shows the method John Canton explained in 1750 to the Royal Society, invited there by Knight before they became bitter rivals. The frilly cuffs shown on Canton's disembodied hands must have reassured nervous gentlemen that it was perfectly appropriate for them to handle everyday objects normally belonging to the servants – coal tongs and pokers, which often became magnetic of their own accord after they had been leaning up against the hearth (in modern terms, magnetism was induced through alignment with the earth's field). First Canton stroked a steel bar repeatedly with his fireside tools until it was magnetised. Next he laid out four other bars on a table, carefully lining them up with respect to the direction of magnetic north, and then embarked on a lengthy reiterative stroking process, swapping bars round until they had all acquired the same strength. His final

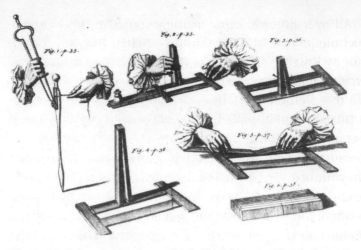

Figure 13: John Canton's method of making artificial magnets. *Philosophical Transactions* 47, 1751, pp. 34–5. (Whipple Library, Cambridge)

illustration shows how several bars could be bundled together to create a powerful battery (electric batteries had not yet been invented).

Canton and Knight had a third English competitor – John Michell, a Yorkshire clergyman. Michell is now famous for his delicate experiments on gravity, for his studies of earthquakes and for (supposedly) proposing the existence of black holes. But in the middle of the 18th century, he was best known for his short book on artificial magnets. The initial comradeship among Canton, Knight and Michell rapidly degenerated into mutual hatred, although Knight finally triumphed as England's leading magnetic expert. Even after Canton and Michell had died, their descendants fought bitterly over who had first invented the 'double touch' method illustrated in Canton's diagrams – using two stroking magnets fastened at a small angle to each other.

All over Europe, experimenters started to try out the techniques published by Canton, Michell and others (but not by Knight!). During the first part of the century, the world's major magnetic research centre had been Leiden in the Netherlands, then a leading university. After artificial magnets were invented, London and Paris both became important, followed closely by St Petersburg. Because the Tsars were keen to modernise Russian science, they funded generous Fellowships which attracted some leading German scholars, including Franz Æpinus, professor of astronomy at Berlin and an expert on electricity. Æpinus developed some of the best techniques for making artificial magnets, and he also wrote a theoretical book that was very influential throughout Europe – although it made less impact in England, because it was too mathematical.

Knight boasted that his magnets were the best available, and this may have been justified at the time. As with many practical tasks, some operators were more skilled than others; in addition, craftsmen failed to publicise valuable tips, such as rubbing the bars with linseed oil, or using particular types of steel. Magnetising bars was a tedious, time-consuming process, and prices reflected the amount of work involved. Cheap magnets – costing only a few pennies were small, weak, wrapped in black paper and made of low-quality metal. In contrast, Knight targeted the top end of the market. He advertised three versions – up to ten guineas for a pair fifteen inches long – all sold in mahogany boxes so beautifully crafted that at first sight they seem impossible to open (Figure 14). To preserve their strength, Knight separated the bars with a piece of wood and placed iron keepers across their ends.

Figure 14: Artificial magnets, probably by Gowin Knight. (Science Museum, London)

Knight's bars were impressively strong, yet he short-circuited the manual labour required by building a magnetic machine – the one that he showed off so proudly to visitors at the British Museum (Figure 15). Presumably this machine had previously been concealed within his study so that he could disappear inside and emerge with a bar magnetised almost instantaneously, as if by magic. It was made of two wheeled magazines, giant batteries each containing 240 magnetic bars, which rapidly magnetised a bar (marked K) fastened between them. In the 19th century, Faraday adapted Knight's equipment to use in his own research projects.

Then as now, some customers instinctively distrusted artificial products, and instrument-makers catered for

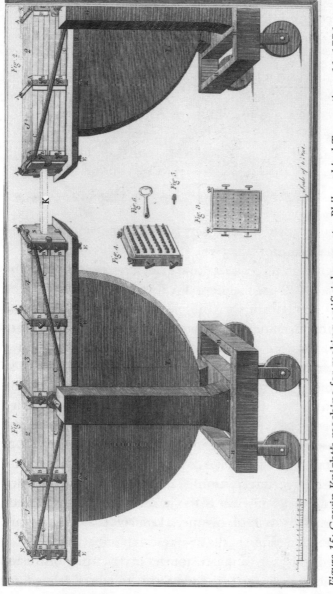

Figure 15: Gowin Knight's machine for making artificial magnets. *Philosophical Transactions 66,* 1776, facing p. 601. (Cambridge University Library)

reactionary tastes by making magnetic bars which looked like loadstones, selling them in fishskin cases lined with velvet. But Knight was proud of his modern invention, and employed no such subterfuge. His priority was to craft portable equipment that would preserve his magnets' strength: he intended them to be bought by serious experimental philosophers, investigators determined to solve the riddle of nature's most mysterious power. Many of these magnetic researchers were – like Knight – scientific entrepreneurs as interested in improving their finances as their expertise.

* * * * *

Because artificial magnets were manufactured under human control, they were far more versatile and consistent than pieces of natural loadstone dug up out of the earth. Of all the magnetic inventions that appeared, the most flamboyant was Temple Henry Croker's perpetual motion machine, shown as Figure 16. Like Knight, Croker was an Oxford graduate, although he was far less skilled at survival strategies. His chequered career included claiming Italian translations as his own, co-authoring a dictionary (mainly a compilation of plagiarisms), declaring bankruptcy and emigrating to the West Indies. Beneath his diagram, the coat-of-arms of Abraham Mason, a settler in Barbados, shows a hand triumphantly brandishing a magnet which supports the weight of a key; *secreta retexit*, proclaims the motto – let the mystery be unravelled – as if this magnetic key would unlock the very secrets of nature.

Croker soon took advantage of the unprecedented possibilities opened up by the new magnetic invention.

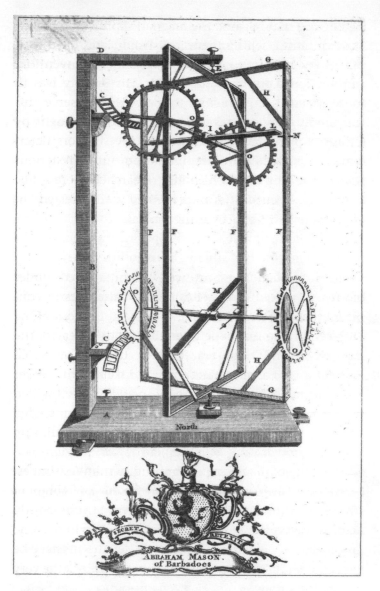

Figure 16: Temple Henry Croker's magnetic perpetual motion machine. Frontispiece of T.H. Croker's *Experimental Magnetism*, London, 1761 © British Library

He decided that because the angle of dip is zero near the Equator, the earth's magnetism would constantly flip over a bar magnet as gravity tipped it downwards. He commissioned one of London's finest instrument makers to construct the extravagant assembly of gear wheels surrounding the central magnet M, and on at least two occasions presented it at the Royal Society – even having dinner afterwards with Knight and some other Fellows at the nearby Mitre inn. Although his machine failed to keep going in London, Croker insisted that it had worked properly in Barbados. This might seem a ludicrous claim, but the Fellows took it seriously because the possibility of perpetual motion had still not been completely ruled out.

The introduction of artificial magnets also revived interest in an old topic – finding how the strength of a magnet falls off with distance. During the first half of the 18th century, several experimenters had tried to convince themselves and their readers that their observations showed a neat mathematical relationship. Their results were inevitably unclear because – in addition to the lack of accurate measuring devices – they had no way of isolating a single magnetic pole. Newton himself had vacillated, bracketing together gravity and magnetism when it suited his arguments, but at other times declaring (without evidence) that there was an inverse cube law (that is, that the magnetic attraction between two objects decreases with the cube of the distance between them). Like many of Newton's followers, Michell was determined to demonstrate that gravity's inverse square relationship held true for magnetism. He even announced that he had succeeded – although there was a suspicious absence of experimental data in his book.

After this brief flurry of investigation, English researchers turned their attention to the earth's magnetic patterns, and it was only decades later – in 1785 – that the French experimenter Charles-Augustin Coulomb eventually persuaded his experimental apparatus to yield the much-hoped-for inverse square relationship. Coulomb possessed a huge advantage over his predecessors – long thin artificial magnets, made from excellent steel according to Æpinus's instructions. Two feet in length, they behaved as though all their magnetic power were concentrated near the tips. In other words, each one was the next best thing to two isolated poles.

Determined to vindicate some of Æpinus's ideas, Coulomb built an extremely sensitive piece of apparatus called a torsion balance (Figure 17). Although the basic principle was simple, making it work accurately was extremely difficult, but Coulomb benefited from years of experience manipulating his delicate equipment to produce precise results. He suspended one magnet (AB) horizontally with a fine wire so that it was free to rotate near his second magnet, which was placed vertically. When Coulomb turned his hanging magnet so that its string was twisted, the angle at which it came to rest depended on several forces affecting it – the torsion in the string, the earth's magnetism, and the attraction and repulsion between the four poles of the two magnets. By repeating the experiment under different starting conditions, and carrying out some mathematical calculations, Coulomb showed that single magnetic poles do, indeed, obey an inverse square law.

It was no accident that this experiment took place in Paris rather than London. The French state poured money

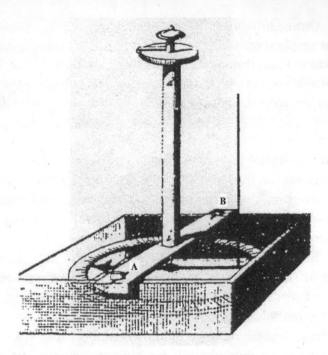

Figure 17: Charles-Augustin Coulomb's magnetic torsion balance. *Histoire de l'Académie Royale des Sciences*, 1785, Figure 4, Plate XIV, facing p. 610.

into scientific research, and the educational system was strongly geared towards engineering and mathematics. But in England, the Royal Society was self-financing, and natural philosophers were encouraged to concentrate on observation rather than calculation. Brought up to study classics and the Bible, many of them distrusted fancy French algebra. Knight's friend Wilson told Æpinus why his theories were unpopular in England: 'The introducing of algebra in experimental philosophy, is very much laid aside with us, as few people understand it; and those who do, rather chose [*sic*] to avoid that close kind of attention.'[12]

Unlike in France, there was no established curriculum in England for students of science, and so experimenters had to learn through experience. And with little prospect of a salaried career, they focused on research projects that might make them some money.

* * * * *

At the same time as Wedgwood was generating new outlets for his pottery, natural philosophers were converting their scientific expertise into products that could be sold. By the end of the 18th century, ambitious parents were convinced that their children – girls as well as boys – could be called well-educated only if they knew some science. The market was soon flooded with books, demonstration equipment and instruments for 'rational entertainment' (an inspired euphemism for conjuring tricks). Almost a third of the 1797 *Encyclopædia Britannica* entry on magnetism was devoted to describing instruments for parlour games: these expensive toys were being cleverly advertised under the guise of education (Figure 18).

Commercialised science boomed, and women as well as men joined in. Although only men wrote specialised texts on magnetism, female teachers were setting up schools and providing chatty introductory books for children. One headmistress, Margaret Bryan, taught physics (including magnetism) as well as publishing books, and Faraday – Victorian England's magnetic expert – first discovered his enthusiasm for science when he read Jane Marcet's *Conversations on Chemistry*. Women were also in the audiences at scientific lectures.

Unlike electricity, whose sparks and shocks could make

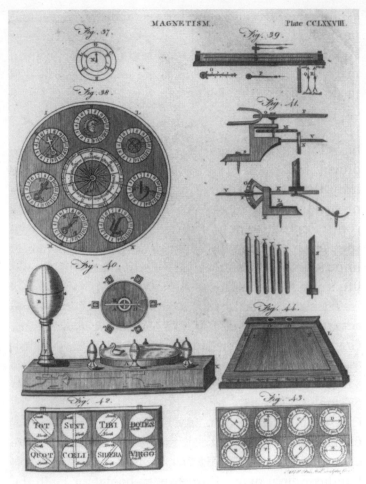

Figure 18: Magnetic rational recreations. *Encyclopædia Britannica*, 1797, vol. 10, facing p. 448. (Whipple Library, Cambridge)

a show literally stunning, magnetism was not intrinsically appealing. Iron filings and compass needles may yield valuable information about the cosmos, but they are not particularly fascinating to watch. Some performers – such as the notorious Gustavus Katterfelto – solved this problem by devising dramatic demonstrations. The self-styled Dr Katterfelto excelled at bringing out the magic in magnetism. Often denounced as a quack because he had made a quick fortune by peddling medicine during a flu epidemic, Katterfelto enjoyed a brief season of fame in London and then trailed round England with his wife, black cat and several children trying to scrape a living in provincial theatres. His publicity material promised some intriguing displays – a 'magnetical Clock' and a mysterious 'Magnetical APPARATUS Which will take a copy off in five minutes time'. One of Katterfelto's most spectacular effects involved strapping his small daughter into a steel helmet so that a giant artificial magnet could lift her up off the ground.

Katterfelto's great rival was James Graham, whose advertising campaign also stressed magnetic mystique. Graham was London's most notorious sex therapist, assisted in his Temple of Health by beautiful Vestal Virgins such as Emma Hamilton (who later became Lord Nelson's lover). His most enticing attraction was his Celestial Bed, which guaranteed (for the enormous sum of £50) a night of connubial bliss while couples were stimulated by the tiny particles circulating from fifteen hundredweight of artificial magnets hidden beneath the mattress.

More sober lecturers faced a dilemma: although they needed to make their performances as attractive (and profitable) as those of Katterfelto and Graham, they also wanted to avoid being bracketed with conjurers and charlatans. But a serious discussion of magnetic poles or navigational compasses was hardly likely to hold their audience's attention. Their solution was to invent classy tricks that seemed to be educational. The most important instruction book for magnetic performers had the clever title of *Rational Recreations*, and it included 57 magnetic tricks.

Many of these rational recreations – enchanted ewers, communicative crowns, dextrous painters – were copied into the *Encyclopædia Britannica*. In the plate reproduced as Figure 18, the bottom two illustrations reveal how small compass needles can be magnetically controlled to display Latin mottoes extolling the virtues of innocent young women. The magnetic planetarium (shown at the top left) was extremely popular: when the central pointer is set to a question, seven small compass needles rotate to give a suitable reply. The other diagrams show the complicated clockwork that lies concealed beneath a favourite French device – the sagacious swan. Guided by the operator's hidden magnet, a small mechanical swan swims round a pool of water to spell out suggestive answers to well-phrased questions.

Magnetic performances did not always run smoothly. A conjuror at London's Haymarket theatre was disconcerted one night when his swan refused to obey instructions. The saboteur wielding his own concealed magnet was delighted with the effect he achieved – the audience started laughing at rather than with the performer. This staged contest symbolises how natural philosophers wanted to distance themselves from showmen: by undermining the conjuror, the educated gentleman in the audience demonstrated his social as well as his magnetic superiority.

In public, natural philosophers behaved with propriety.

They reserved magnetic tricks for the privacy of their own homes – Charles Darwin's grandfather Erasmus livened up his dinner parties by making a small magnetic spider scuttle across a silver salver (even Knight attended a similar event, although there is no record of whether he laughed). But natural philosophers banned such light-heartedness from serious science, because they were trying to set up a firm boundary between scientific lecturers and popular conjurors, between elite experts and untrustworthy charlatans. In reality, there was often little to distinguish them. Rather than imagining two separate groups, it is more helpful to think of magnetic educators and entertainers being laid out along a continuous spectrum of performers, all competing for their customers' money.

To appreciate this situation from another angle, take a look inside George Adams' portable cabinet (Figure 19), complete with its own ingenious magnetic padlock (centre). For imparting a sound magnetic education, this kit contains everything a travelling lecturer or home enthusiast might desire: artificial magnets in two shapes – straight bars (bottom right) and horse-shoes; compasses and dip-needles with detachable stands (top left); and two old-fashioned loadstones in protective casings (centre left). But what about the three hearts at the bottom left? They are intended for entertainment rather than straight education. In this popular trick, hidden magnets deceptively make it appear that a compass needle can pick out different metals.

Adams' fine wooden box also includes a prominent circular map, pasted on to a wooden box with a small magnet inside. This is designed to show students how

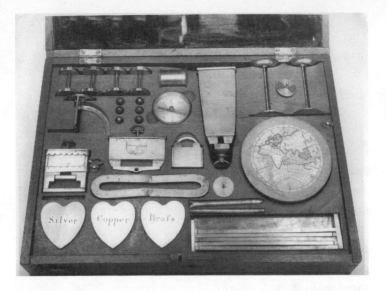

Figure 19: George Adams' magnetic cabinet. (Science Museum, London)

needles behave as they move around the globe. Adams marketed Knight's magnetic instruments, but both men knew where the greatest profit lay – in navigation. And they had a great sales pitch: compasses would help Britain to rule the world.

· CHAPTER 7 ·
COMMERCIALISING COMPASSES

Hail glorious gift! design'd the world to bless,
Transcended only by the teeming Press.
The fearless pilot led by thee shall brave
The turbid fury of the Atlantic wave.

Joseph Cottle, *Science revived, or the vision of*
Alfred, 1801

British maps displayed an important warning for unwary navigators – the site of Sir Cloudesley Shovell's spectacular shipwreck off the Cornish coast in 1707. Shovell had been aiming for the Bristol Channel, but he ran aground far to the south on the Scilly Isles, losing four ships and 2,000 men. Although his casualty figures were unusually high, it was not uncommon for captains to be many miles off target. Add in pirates, bad weather and scurvy, and it becomes clear why Samuel Johnson quipped that 'being in a ship is being in a jail, with the chance of being drowned'.[13]

The official enquiry discovered that although Shovell's squadron owned 112 compasses, only four of them had been properly maintained. The accident was not, it seemed, entirely due to Shovell's incompetence: false economies and administrative errors had also contributed to this legendary loss. In response, naval reformers welcomed suggestions for improvement, and the government promised enormous rewards for solving the longitude problem.

Armchair travellers waxed lyrical about the benefits of magnetic compasses. Loadstone might appear to be a 'mean, contemptible and otherwise worthless *fossil*', remarked the religious poet James Hervey; but in reality, he explained, it is God's great gift to the British nation. A cheap compass, he enthused, 'is the means of conveying into our harbours the rarities and riches of the universe ... the choice productions, and the peculiar treasures, of every nation under heaven'.[14] Like Hervey, sedentary missionaries ensconced in the comfort of their own studies preached that compasses enabled 'the industrious bees, from the hives of *Europe*' to sail to the other side of the world and instruct 'the *Indian* and *African* savages ... in the knowledge of that supreme Lord and Governor of the universe, of whom, before that, they had such odd and uncouthly confused notions'.[15]

Experienced sailors were more sceptical. Many tars – seasoned seamen – preferred to rely on traditional methods of navigation rather than place their trust in unreliable instruments. Compensating for magnetic variation involved long calculations – and mistakes could cost lives. Navigators were more interested in arriving safely than in pinpointing their position on an inaccurate chart. Relying on traditional methods of steering by landmarks, they taught their recruits 'that a learner will steer a ship to a greater nicety by a mark-a-head, than a good helmsman can do without a mark by the compass'.[16]

Shovell's crews were using compasses whose design changed little over a couple of hundred years. The French example shown in Figure 20 is typical of the 17th century, even though it dates from the 1770s. The needle is invisible because it is attached beneath the card carrying

Figure 20: French compass by Joseph Roux of Marseilles,
c. 1775. (Science Museum, London)

the decorative rose. This rose is designed with traditional
symbols – a fleur-de-lys for north, letters for Italian winds
(for example, west is P for *penete*), and a cross for east, the
direction of both Jerusalem and Paradise; north is now
usually at the top of maps, but this convention was not
always followed then. The word 'needle' is misleading,
because this one is made from a length of soft iron wire
bent into a crude diamond shape; inherently imprecise, it
also needs to be regularly remagnetised with a piece of
loadstone. The card's weight means that the compass
responds only sluggishly, and the wooden bowl has
swelled and warped from being soaked in salty spray.
Because the glass is fixed in with putty it was not replaced
when it cracked – another opportunity for water to seep
in, saturating the card and rusting the needle.

Campaigners recognised that although magnetic navigation could be improved by providing better compasses, reform would also involve changing the behaviour of sceptical mariners. One problem was ignorance. Garlic and onions were banned on many ships because of their (supposed) magnetic effects, and even senior men failed to realise that iron would distort the compass readings. Binnacles – the protective cupboards for compasses – provided convenient storage places for pistols and muskets, and James Cook stored the keys to his leg irons next to his compasses. Naturally, Cook blamed the compasses rather than himself. 'I do not remember of ever finding two needles [compasses] that would agree exactly together at one and the same time and place', he complained.[17]

Another difficulty lay in the way compasses were used. New technologies are often taken up in ways undreamt of by their inventors. Mobile phones, for instance, were initially marketed to rich businessmen as a tool for urgent conversations, but are now bought by teenagers who constantly exchange trivial text messages. Similarly, sailors adapted for their own purposes the magnetic instruments introduced by land-based entrepreneurs. They lined up compasses with a ship's wake to find its leeway (the angle between the keel and the actual course); and to compensate for variation, they often rotated the compass card. In Figure 20, the lubber line marked on the inside of the bowl shows that this card has been adjusted through 20°; this practice worked well for local voyages, but gave faulty readings if the ship ventured into an area with a different variation.

Naval administrators came in for a good deal of

criticism. They were frequently accused of bureaucratic bungling – providing insufficient training, skimping on provisions, short-circuiting test procedures. The hand-written minutes of Admiralty committee meetings reveal that the chronic shortage of funds was often resolved by 'the Sale of decay'd Naval Stores'. As the backlog of unrepaired compasses piled up in the dockyards, many experts agreed that 'the mechanical parts of our com-passes, too, are generally, very indifferently executed; indeed, the low price allowed by the government for supplying the navy with that instrument will not admit of their being made by the best workmen and in the most correct manner'.[18]

Inaccurate compasses, superstitious sailors, incom-petent administrators – an ideal opportunity for Knight to step in as scientific hero and resolve the situation. Shipping was vitally important in the 18th century, and the Fellows of the Royal Society stressed the commercial benefits of Knight's revolutionary compass design. Thanks to Knight, bragged President Folkes, 'we have been enabled to make, with security and ease, long voyages by sea; and consequently to increase and promote greatly our foreign trade and commerce, whereby we are provided at home with the fruits, the conveniences, the curiosities and the riches of the most distant climates'.[19]

But introducing new technology does not necessarily result in progress. The real story was more complicated. Teething problems turned out to be serious, and it was only in the middle of the 19th century that new solutions were found to some old problems. Knight did radically overhaul the design of compasses, and he did produce sensitive instruments that gave very accurate readings.

But he never sailed across an ocean. And as any mariner could have told this landlubber, 'there is no such thing as Preciseness to be expected from any Mathematical Sea Instrument whatever, as most of them are liable to Error from the Motion of the Ship'.[20]

* * * * *

In 1749, the crew aboard a ship sailing from New York to London had a frightening experience. A five-day storm blew up, with thunder 'and sundry very large Comazants (as we call them) over-head, some of which settled on the Spintles at the Top-mast Heads, which burnt like very large Torches'. These abnormal lightning flashes, now known as St Elmo's fire, 'drew the Virtue of the Loadstone from all the Compasses … they were at first very near revers'd, the North to the South; and after a little while rambled about so as to be of no service'.[21]

The Royal Society invited its magnetic expert – Gowin Knight – to examine one of these damaged compasses, and he lost no opportunity of displaying his expertise to the Fellows. Although many of them probably dozed off during the technical explanations, his central message got through. 'It will cost only about 2s.6d. more to buy a tolerable good [compass],' one listener recorded, 'So that the Lives and Fortunes of thousands are every Day hazarded for such a trifling Consideration.'[22] Knight had made it clear that his scientific innovations would save money and lives – if only the Navy could be persuaded to make the initial investment.

After investigating many compasses, Knight introduced major changes in the design of the casing and the needle (Figure 21). He had been horrified to discover that

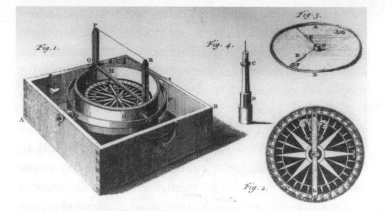

Figure 21: Gowin Knight and John Smeaton's azimuth compass. *Philosophical Transactions* 46, 1750, facing p. 515. (Whipple Library, Cambridge)

compass cases were fastened together with iron nails, and he insisted that his compass be mounted in a high-quality brass bowl with a hinged glass lid; to compensate for the ship's motion, Knight designed an adjustable suspension system to support the bowl inside a wooden box made with brass screws. These were costly and labour-intensive changes: non-magnetic brass was expensive, and each screw was hand-cut, so would fit only its own hole.

To find the shape of needle that would give the most regular pattern, Knight carried out experiments with scatterings of magnetic sand. He decided to use a long, narrow rectangular bar made of high-quality hard steel which would hold its magnetism well and not rust. To avoid piercing it, which might introduce asymmetries, Knight balanced this needle on a fine point above a light card supported by a brass ring round its edge. At this stage, he was investing his own money in his research, and he was determined not to let the expenditure climb too high.

So Knight made his point from an ordinary sewing needle, which could easily be replaced if it became blunt.

Acting as a supportive patron, Knight allowed his assistant Smeaton to describe the azimuth features of their compass to the Royal Society. Smeaton showed the Fellows how, using the narrow sights, an observer could align the compass with a distant object, and then record the position of the shadow cast by the horizontal string onto the face. After some mathematical calculations, this reading would yield the magnetic variation. The diagram on the top right shows two radially adjusted weights used to compensate for dip. Smeaton explained that the card's brass ring was finely marked so that it could be read to less than half a degree. He somehow omitted to point out that the instrument had to be rotated by hand, an operation hard to perform accurately on land, let alone at sea. 'Of all the Instruments now commonly used at Sea', wrote one disillusioned mariner, 'I do not know one from its Construction so clumsy, or in the Use of which People are more imposed on, than that which goes by the Name of the Azimuth Compass.'[23]

Harshly squeezing out Canton and other potential competitors, Knight secured a lucrative contract with the Admiralty and was awarded £300 development funds from the Board of Longitude: the minutes record that this money was to be raised by selling off old equipment. To customers visiting his London instrument-shop, Adams charged high prices for Knight's magnets and compasses, but Knight sold them directly to the Navy at a discount. Other potential purchasers were deterred by the cost, but converts (perhaps encouraged by Knight) stressed the benefits: 'Where there is so many lives, and so much property depending on good Compasses,' exclaimed the Liverpool Dockmaster, 'I have been surprized and vexed to hear some people begrudge the price of Dr. *Knight's* improved steering and azimuth Compasses, which I thought, when I bought one of each, not only deserved the price, but the inventor the thanks of the public as a trading nation and a maritime power for so great an improvement in that important instrument.'[24]

Knight's compasses were soon standard issue for all ships embarking on international voyages. However, the testing programme was perfunctory. Knight and Smeaton did venture a few miles off the English coast, but soon abandoned their trials when 'Fresh gales with strong Squalls and hails pm filled 15 punch[s] with Salt water Opened a Cask of Beef at 6am weighed & Came to Sail lost the Logg & three Lines trying Experiments'.[25] The Admiralty ordered tests on five long-distance ships, but – economising as usual – bought only enough equipment for three. Just one captain sent in his opinion: Knight's compasses are, he reported, 'preferable to the other Compasses now in use at all times, except in Stormy Weather'.[26]

'Except in Stormy Weather' – if only someone had listened at that stage. Initial enthusiasm soon turned to despair as horror stories flooded in. Navigator after navigator complained that because Knight's compass was so sensitive, the needle simply spun round and round in tempestuous waves. When Cook lost his favourite but old-fashioned compass overboard, he demanded an identical replacement and rejected newer models because 'Doctor Knights Stearing Compas's from their quick motion are found to be of very little use on Board small

Vessels at sea.'[27] Like many seamen, Cook preferred the traditional version, whose heavy card stabilised the fluctuating needle.

As a natural philosopher, Knight had focused on extreme precision. In contrast, mariners out at sea were satisfied with making sure of their general direction. One experienced officer approved of Knight's compasses because they were accurate to within a point. But a point is over eleven degrees, and Knight had been aiming at fractions of a degree. Seafarers and natural philosophers used different units. Knight had learnt about circles in his geometry lessons, whereas maritime men thought in terms of wind directions. They divided the circle of the compass face into quarters, and then repeatedly subdivided by halving: what Knight knew as 360 degrees, they called 32 points. Knight and his successors kept introducing modifications, but the complaints still poured in. 'The New Compass is a very fine Instrument,' remarked an explorer about to set off for Tahiti, 'but will not answer at Sea in bad weather when the Ship has a quick motion. It then runs round and never stands Steedy.'[28]

This clash of perspectives between natural philosophers and maritime men was made more severe because information passed only slowly between them. The Royal Society chose to ignore the complaints of sailors, who were often men from a lower social class. Long after Folkes died, Fellows still singled out Knight in order to trumpet their achievements. 'Let those who doubt', one President thundered, 'view the Needle, which, untouched by any loadstone, directs the course of the British mariner round the world.'[29]

As late as the 19th century, when one ship's captain protested that all his compasses were defective, it was his passengers who pointed out that the ship had an iron cargo. Administrators were reluctant to introduce changes, and they continued to make false economies. In 1824, a teacher from the Woolwich Military Academy was scandalised when he inspected some naval stores. 'I could scarce bring myself to believe that the instruments exhibited to me were those actually employed in His Majesty's vessels', he exclaimed; 'it does appear to me very unaccountable that vessels of such immense value, and the safety of so many valuable lives, should be so endangered by the employment of instruments that would have disgraced the arts as they stood at the beginning of the eighteenth century.'[30]

All the protests started to be effective only after the problem had become worse. In the early 19th century, iron ships were introduced, and it became essential to find new ways of making compasses reliable and useful. By then, power structures had also altered. The Royal Society had negotiated itself into a far more prestigious role, and navigators were receiving mathematical education as well as on-the-job training. The Fellows, civil servants and naval officials all recognised that it was in their mutual interests to invest in magnetic navigation, and Societies from different countries also started to collaborate. During this new era of scientific cooperation, international projects of exploration charted the earth's magnetic patterns to unprecedented levels of accuracy. Global Victorian precision took over from local Enlightenment innovation.

· CHAPTER 8 ·
ENIGMATIC THEORIES

'I am not very well versed in Greek', said the giant: 'Nor I neither', replied the philosophical mite. 'Why then do you quote that same Aristotle in Greek?' resumed the Sirian: 'Because', answered the other, 'it is but reasonable we should quote what we do not comprehend in a language we do not understand'.

Voltaire, *Micromégas*, 1752

During the Enlightenment, educated people – including some women – boasted about belonging to the international Republic of Letters, an imaginary community of intellectual correspondents that spread out through Europe and the American colonies. In reality, this ideal Republic was not a united one. Natural philosophers of each country had their own views about which theories to adopt and how experiments should be conducted. They even had their own national champions – Descartes in France, Newton in England, and Leibniz in Germany – whose supporters continued to argue with one another long after these three great figureheads had died.

France and England were often at war with one another, and military hostilities spilled out into an intellectual battlefield. Frenchmen were, sniped one of Newton's henchmen, an 'Army of Goths and Vandals in the Philosophical World'.[31] According to Voltaire, travelling across the Channel meant arriving in a different world. Banished from France for his radical beliefs, Voltaire had retreated to England, where he claimed to have landed in a haven of liberty and equality. After marvelling at the splendour of Newton's funeral procession, Voltaire became an ardent campaigner for everything English, a subversive strategy designed – successfully – to annoy the French authorities.

Voltaire reported that English space was literally empty. 'In Paris they see the universe as composed of vortices of subtle matter, in London they see nothing of the kind', he remarked. 'For your Cartesians everything is moved by an impulsion you don't really understand, for Mr Newton it is by gravitation.' English Newtonians envisaged a universal force of gravity which pulled apples to the ground and reached out between the sun, the planets and distant stars. In contrast, French Cartesians insisted that it was impossible for objects to affect each other across empty space.

In England as well as in France, 17th-century scholars had welcomed Descartes's efforts to rid natural philosophy of occult qualities, those strange hidden forces that allegedly extend throughout the universe. Like him, they wanted to relegate such mysterious phenomena to the realm of magic, and instead produce rational explanations based on common sense and observation. If ordinary things could somehow communicate without any intervening medium, Descartes had protested, then brute matter would have to possess some sort of non-material power. But for devout Christians, such a spiritual quality was reserved for God and human souls. In Descartes's orderly clockwork cosmos, everything happens by direct mechanical contact. Set into motion by God at the creation, it is packed with swirling streams of minute particles which collide to produce visible effects.

Because magnets behave so mysteriously, Descartes realised that it was essential for him to provide a mechanical explanation. Magnetism was his test case: if he failed with this, then it would be easy for his critics to reject his entire model of a cosmos crowded with spinning vortices of particles, and instead retain their traditional beliefs in occult forces. To illustrate his magnetic hypotheses, Descartes used the woodcut reproduced as Figure 22. The large circle in the centre represents the giant loadstone of the earth, with its south pole at A and its north pole at B; the five small ones indicate terrellas placed in various latitudes.

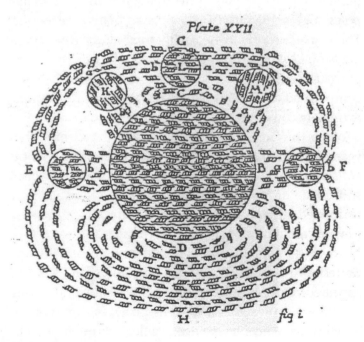

Figure 22: René Descartes's magnetic earth. *Principia Philosophiæ*, Amsterdam, 1644, p. 274.

This diagram may resemble a modern magnetic field, but the underlying theory is completely different. According to Descartes, the two poles of a loadstone are connected together by long channels bored out with a corkscrew lining. Streaming through these pores are magnetic particles or effluvia, which are characterised by a screw thread allowing them to travel in one direction only. Particles continually pour out of one pole, then circulate round the earth to enter the opposite pole at the other end. To make matters more complicated, Descartes suggested that there are two different kinds of particle, screw-threaded in opposite directions. His model works quite well in its simplest form. Imagine bringing two like poles together with effluvia pouring out of each one. These particles will be unable to enter the other loadstone's channels, and so the two loadstones will be pushed apart by a cloud of colliding effluvia. But for more complicated situations, Descartes's accounts became extremely convoluted.

When Descartes died in 1650, Newton was only seven years old. Almost 40 years later, when he published his *Philosophiæ Naturalis Principia Mathematica*, Cartesianism still ruled Europe. It was no accident that Newton's title resembled that of Descartes's *Principia Philosophiæ*: Newton aimed to replace the Frenchman's system with his own. In romanticised versions of science's history, Cartesianism was abandoned as soon as Newton appeared. This mythical view of events was summarised by the poet Alexander Pope. His pithy epithet glorifies Newton as God's messenger, whose inspired genius instantaneously transformed human understanding of His universe:

Nature and Nature's Laws lay hid in Night,
God said, *Let Newton be!* and all was Light!

Reality was different. Change was far slower on both sides of the Channel. The French education system was dominated by Cartesian Jesuits who effectively excluded Newtonian gravity until the second half of the 18th century. In England, after much initial resistance, natural philosophers generally adopted most parts of Newton's ideas, rejected some entirely, and combined others with their own insights as well as with odd bits of Cartesianism that they liked. To consolidate this Newtonian hegemony in England, propagandists censored Cartesian textbooks when they were translated into English, adding pages of footnotes pointing out why Descartes was wrong and Newton was right.

Magnetism was the big exception to Newtonian domination: the industrious editors left pages of Cartesian magnetism unscathed. Natural philosophers wanted theories that worked, and although Descartes's explanations often limped, they did provide answers to some questions – especially magnetic ones. Newton himself had written very little about magnetism, and English natural philosophers set Gilbert up as their magnetic hero. However, their theories lay far closer to those of Descartes.

During the 18th century, English experimenters could find virtually no information about magnetism in English textbooks, and so they turned to French ones. Faced with the lack of any viable alternative theory, they glossed over the fine details of Descartes's theories and assumed – like Halley – that streams of effluvia are coursing through the invisible pores of iron and loadstone. After all, as an eminent London physician explained, you can almost see the invisible particles in action: 'how instantaneously the *Effluvia* of a Loadstone affect the Fileings of Iron laid on a Table; that it will make them all stand upright in a Moment, and if you withdraw the Stone, or intercept the *Effluvia* with an Iron Plate, that these Effects cease as soon'.[32]

But staunch Newtonians continued to regard Descartes as the enemy. Loyal sceptics chauvinistically dismissed magnetic effluvia as 'that Romantick Solution of the *French* Philosopher'.[33] Officially, Cartesian magnetism was rejected. However, lightly disguised, it did creep into Newtonian philosophy.

* * * * *

During the whole of the 18th century, only three Newtonian philosophers wrote books on magnetism: William Whiston (1721), Gowin Knight (1748) and Tiberius Cavallo (1787). And the only author who tried to develop a comprehensive theory was Knight.

Once you venture past the title, you soon realise that his *An Attempt to Demonstrate, that all the Phænomena in Nature may be Explained by Two Simple Active Principles, Attraction and Repulsion: Wherein the Attractions of Cohesion, Gravity, and Magnetism, are Shewn to be One and the Same; and the Phænomena of the Latter are more particularly Explained* is not going to be a gripping read. Knight was not a man to write in sound bites, and although his colleagues often mentioned his book, they refrained from quoting it. Knight's description of two north poles repelling one another suggests one reason why: 'there must be a Conflux of repellent Matter towards both Poles to supply their Streams, which must make a double Flux of it towards the intermediate Space; and because both the North Poles are supplied by Streams coming from their contrary South Poles, the Stream coming to the one will be opposite to that coming to the other.'[34]

Despite being so opaque, this quotation does make it clear that Knight's conjectures owe a good deal to Descartes's. No effluvia appear in his book, but conveniently vague terms such as 'fluid' and 'virtue' abound. 'Fluid' was a particularly useful word for Knight because it could mean either a gas or a liquid, and so might reasonably be expected to have some strange properties. Nevertheless, Knight's role model was Newton. He even made sure that his own book resembled Newton's *Principia* – the same quarto size, well produced, and neatly laid out like a geometrical argument in numbered propositions and corollaries. Unlike Newton, however, Knight was not an expert mathematician, and there are very few diagrams and no equations in his book. The nearest he came to maths was his ambiguous phrase, 'the Quantity of relative Magnetism'.

Knight presumably scoured Newton's work for guidance. He would have found only a few references, none of them very helpful. When it suited his purposes, Newton chose to assume that magnetic effluvia really do exist. So to support his thesis that things which appear to be solid are actually porous, Newton wrote: 'Gold is so rare as very readily and without the least opposition to transmit the magnetick Effluvia.' This declaration made an easy target for his opponents. In that case, they demanded to know, why is gold not transparent to effluvia of light? And why could magnetic effluvia not pass through iron? At other times, Newton opted to take it for granted that magnetic activity is due to forces acting across space. 'Have not the small Particles of Bodies certain Powers, Virtues, or Forces, by which they act at a distance', he asked. Without pausing for an answer, he continued: 'For it's well known, that Bodies act one upon another by the Attractions of Gravity, Magnetism, and Electricity.'[35]

Newton's critics savaged his notion that attraction operates through empty space. They denounced him for bringing back occult qualities under another name, and challenged him 'to shew us the Chain which fastens ... Iron to the Loadstone'. English Cartesians who refused to be converted accused Newton of circular arguments: 'little are wee helped by the late word In use, attraction; for that is Idem per Idem. and If one asks why one thing draws an other – It is answered by a certain drawingness it hath.'[36] To silence these sceptics, Newton suggested – with no great precision – that there might be some sort of subtle aether which permeates the entire universe and transmits light and gravity, electricity and magnetism. Particles could, he proposed vaguely, both attract and repel other particles, depending on their size.

Newton's disciples gradually started to explore these radical hypotheses, and by the time that Knight came to write his own book, other natural philosophers had developed mathematical and experimental reinforcements of Newton's aether. Knight's basic assumption was a Newtonian one: that matter is fundamentally composed of minute particles which are either attractive or repulsive. Knight envisaged these tiny atoms clustering round one another to build up larger corpuscles of various sizes, whose attractive or repulsive properties depend on how many of each basic particle they include. These positive and negative corpuscles then combine together to produce the different gases, liquids and solids which make up the world around us – and also ourselves.

Armed with this blueprint of nature, Knight set about reconstructing the entire edifice of physics. He devoted the last third of his book to magnetism, which he attributed to a fluid of mutually repellent corpuscles flowing through the pores of solid materials. Knight had undertaken many meticulous experiments, watching carefully to see how his iron filings aligned themselves in various situations. Yet in describing his observations, he constantly slipped verbally between the filings that he could see and his streams of fluid that were hypothetical. Even in his first magnetic explanation – Corollary I of Proposition LXXVII – Knight retreated into assumptions rather than arguments: 'For the Manner in which Steel-Filings dispose themselves round a Loadstone expresses very exactly the Direction of the magnetic Virtue, and very well corresponds with the Circulation ... of the repellent Fluid.'[37]

For page after page, Knight regaled his readers with tedious accounts of iron filings being marshalled into order by his circulating fluid. Fluid leaving similar poles pushes them apart, and fluid sweeping round loadstones pushes unlike poles together. Knight would probably have been horrified at any delicate hint about a similarity between his fluid and Cartesian effluvia. In his defence, he could have pointed to some important differences: Descartes envisaged counterflowing streams of different particles, whereas Knight had a single repellent fluid; Descartes's account was qualitative, whereas Knight scattered references to inverse square relationships throughout his text. But with hindsight the resemblances seem obvious.

Knight was on stronger ground when he relied on his own unique experiences. Because of the research time he had invested in producing artificial magnets, Knight was – unusually for a Fellow of the Royal Society – familiar with techniques of ore refining and steel manufacture. He drew directly on the studies of two foreign natural philosophers who were, like him, engaged in practical work: the metallurgical investigations carried out by his French counterpart at the Paris Académie, René Réaumur; and the phlogiston theories developed by the German chemist Georg Stahl.

Réaumur had aimed to demonstrate that natural philosophy could have industrial applications. He produced detailed microscopic descriptions of various metals, and Knight used Réaumur's analyses to account for the contrasting magnetic properties of iron and steel. The two metals had, Knight explained, very different sizes and shapes of internal pores for his repellent fluid to stream through.

Réaumur made the internal pores of metals visible, whereas Stahl never claimed to have seen phlogiston, the hypothetical substance that he introduced. Stahl's goal was to understand the smelting processes that miners had been using for centuries. Why is it, he wondered, that the best way of obtaining pure metal is to heat its ore with charcoal? Stahl provided an answer: charcoal is full of phlogiston, an invisible, colourless and weightless fluid which is absorbed by the ore. And he also watched the reverse process taking place: when charcoal is burnt, only a few ashes are left behind because the phlogiston has escaped.

Now that oxygen has been introduced into chemistry, phlogiston seems ridiculous. But it did explain some reactions very satisfactorily, and Knight is an exceptionally early example of its English advocates. For Knight, phlogiston was not an imaginary subtle fluid, but was real, visible, and made metals look black. 'That there is in Steel a very large Quantity of the *Phlogiston*, is beyond Doubt', he wrote; 'soft Steel has a dusky appearance; so that we see the *Phlogiston* in its Pores with the naked Eye.' In contrast, he explained, hard steel lacks phlogiston until it is heated, when 'it comes out again into the visible Pores, as the Steel grows hot; and even spreads itself all over the outward Surface'.[38]

Knight was pompous and verbose, but he was not stupid. Repellent fluids and phlogiston seem zany now, but in the absence of any better suggestion, they did provide helpful models. Enlightenment writers liked metaphors of circulation. They boasted that an enlightened English physician – William Harvey – had overturned centuries of ignorance by demonstrating how blood circulates round the body. Similarly, they argued, fresh air should circulate through a room to guarantee good health, labour and goods should circulate freely to make the economy flourish, and traffic can circulate efficiently only through a well-designed city. It made sense to imagine invisible magnetic fluids circulating around the earth, cascading through the atmosphere as they magnetised iron window bars and looped through the pores of loadstones.

* * * * *

Even before the French Revolution of 1789, Francophobia was widespread in English scientific circles. Knight's successor as magnetic expert at the Royal Society was Tiberius Cavallo, an Italian immigrant who was one of the few Fellows to be enthusiastic about the new French balloons, first successfully launched in 1783. Brave (foolhardy?) inventors were staging spectacular flights which attracted large paying audiences. Sober English natural philosophers disdained such showmanship, but Cavallo took balloons seriously: like French investigators, he recognised the research potential of carrying measuring equipment high up into the atmosphere.

Cavallo worried that his Catholicism made him an outsider, but he had certainly acquired an English scepticism towards grand theory. Forty years after Knight, Cavallo published his own *Treatise of Magnetism*, and he made his position clear from the outset. He denounced the majority of magnetic theories as rubbish (naturally dressing up this verdict in polite euphemisms). Instead of banging his head against the wall of the unknowable, he would, he announced, concentrate on facts and laws, on observations and applications.

As part of their anti-Cartesian rhetoric, Newtonians condemned abstract hypotheses: they insisted that the only true route to knowledge lies via experimental research. In addition, British natural philosophers were inclined to focus on practical topics because they needed to earn money. One of Samuel Johnson's Enlightenment biographers commented approvingly of the situation in France where 'genius was cultivated, refined, and encouraged'. In contrast, he remarked disparagingly, the English public was 'engrossed by ... trade and commerce, and the arts of accumulating wealth'.[39] Unlike their counterparts in continental Europe, British researchers received no state funding, and – like Knight – they wanted to show how useful their investigations could be.

Theories about magnetism were not just about the physical world: they were also about religious beliefs. Magnetic aethers proliferated during the 18th century, but although they often sounded very similar to one another, they were devised for very different theological purposes. For many Newtonians, magnetic fluids provided some sort of intermediary that enabled God to interact with ordinary matter – they were 'the universal Mover of all gross Bodies, and the immediate Cause, under God, of all natural Actions'.[40] But pious philosophers clung to literal interpretations of the Bible. They rejected attraction at a distance because they denied that matter can be intrinsically active, and they envisaged the universe as God's great machine driven by divine etherial streams circulating around the sun. For them, magnetism was a type of holy spirit, God's direct agent in the universe. One inspired surgeon wrote that a magnetised steel bar is laden with activating fire like a budding elder stick primed with life.

Some English philosophers maintained that it was sacrilegious to search too closely for the causes of magnetism. God had intentionally made magnetism mysterious, they claimed, to teach us modesty and force us to recognise how little we can ever know. This argument became particularly useful for conservatives wanting to attack French thinkers after the Revolution. Knight's instrument maker George Adams had political radicals in mind when he challenged 'the modern philosopher, who denies the existence of a God [to] tell you what magnetism is, and how it exists'.[41]

This aversion to theory and to France also entailed strong resistance to mathematics – especially the new French algebra. British natural philosophers were suspicious of abstract symbols, sneering that algebraists performed like bank clerks, mindlessly manipulating letters which were disconnected from the phenomena of the real world. They believed that advanced mathematics had only two uses – practical applications and sharpening minds. Chauvinistic writers denigrated algebra and Frenchmen with the same pejorative adjectives – flowery, superficial, ostentatious – and preached that over-indulgence in fancy algebraic sophistication would distract students from more worthwhile subjects, such as classics and the Bible.

So when Knight's friend Wilson spurned Æpinus's models of magnetism and electricity, his main message was that he and his colleagues distrusted algebra. Wilson himself was probably not very good at mathematics, but he did not simply mean that no British people were. After all, rejection need not stem from incompetence. One of the most ardent anti-French campaigners was John Robison, a Scottish professor of physics who certainly understood Æpinus's algebraic arguments, but maintained that geometry, not algebra, was the right way to approach magnetic problems.

Towards the end of the 18th century, Robison changed the course of theoretical magnetism. Although Æpinus had made magnetism mathematical, his theory still involved a magnetic fluid. It had obvious roots in Cartesian ideas – this was one of the reasons why the Parisian Coulomb admired Æpinus so strongly. In contrast, Robison spurned invisible fluids and aethers, and campaigned for a return to Gilbert: he insisted that close observation was the key to success. Robison reinterpreted the patterns of iron filings round a bar magnet, arguing that each filing itself becomes a small magnet; these small individual magnets then align themselves in chains around the larger magnet. Using diagrams like the one in the top left of Figure 9 on page 78 (from Thomas Young's lectures at the Royal Institution), Robison described the curves by analysing the filings' mutual interactions geometrically.

Robison is often portrayed as a paranoid crank because he became a conspiracy theorist convinced that masonic revolutionaries were taking over the world. Robison even regarded the *Encyclopædia Britannica* as a national defence system against the perfidious propaganda purveyed in the French *Encyclopédie*. But although his chauvinism was extreme (the polite way of putting it – historians are always loath to label someone as mad), everyone knew that there were significant regional differences even inside Britain. Robison had been trained in the Scottish common sense school of thought: he accepted as true whatever was agreed upon by rational men (women didn't count in this argument because they were irrational by nature). With his emphasis on geometry and careful observation, Robison introduced English men of science to a Scottish way of thinking. His geometrical analyses of the tiny filings curving round magnets laid the foundations for Faraday's field theory and Victorian electromagnetism.

PART THREE
MESMER'S MAGNETIC
MEDICINE

I ask'd Mrs Sheridan whether there was any truth in what We had heard concerning her having experienced the effects of [Animal Magnetism]. She confess'd that she had and ... The Duchess of Devonshire was thrown into Hysterics, Lady Salisbury put to sleep the same morning – And the Prince of Wales so near fainting that he turned quite pale and was forced to be supported.

Betsy Sheridan (the playwright's sister), *Journal*,
October 1788

February 1784, London
Tiberius Cavallo (or Don Tiberio as he is known to his friends) is making only slow progress with his book on magnetism. He promised to finish it last November, but – against his better judgement – he keeps getting diverted by balloons.

Spring 1785, London
Putting magnetism temporarily to one side, Cavallo has published his *History and Practice of Aerostation*. Two intrepid pilots have recently crossed the Channel from Dover to France, and spectacular flights are drawing huge audiences all over England. Could this turn out to be the big opportunity Cavallo has been hoping for?

Summer 1785, Paris

As a carnival winds its way through the narrow streets, miniature balloons float above a line of jesters who seem to be suffering from convulsive fits. The crowds jeer at an ass which has been dressed up to represent Franz Mesmer, the renowned but discredited purveyor of magnetic medicine. Peddlers hawk cheap prints of a caricature showing Mesmer drifting off to nowhere in a balloon. 'Adieu, baquet, vendanges sont faites', reads the caption – 'Farewell, magnetic tub, the rip-off is over.' One scientific fashion has been replaced by another.

Summer 1786, London

Mesmer has been banned from Paris, but a few of his followers are setting up private practices in England. The President of the Royal Society has condemned 'Ballomania', and flying is already becoming passé in England; aeronautical spoofs are, however, selling well. Cavallo has abandoned ballooning and gone back to magnetism. He writes some scathing comments on magnetic therapies but publishes nothing about Mesmer because he is in an awkward position – one of his close friends is an enthusiast.

Summer 2004, Cambridge

A historian of science sits wondering how to include Mesmer in her own book about magnetism. Like Cavallo, she tends to get sidetracked, and she starts musing about a children's book she had once read aloud at bedtime. 'The cruel Roman emperor mesmerised the beautiful princess', she had found herself saying – but decided not to prolong progress towards sleep by explaining that the word was being used anachronistically.

Daydreaming on, she reflects that although the verb 'to mesmerise' has been in the English language for 200 years, few people are aware of its origins. And even those who have heard of Mesmer believe that he was an ignorant quack. Events of the past are always being reinterpreted, and for the past 40 years, medical historians have been creating different explanations of why Mesmer was banished from Paris and excluded from orthodox medicine.

She decides she will start by summarising the conventional story – Mesmer as charlatan. And then, still sticking closely to the historical facts, she will subvert this traditional tale of an opportunistic con-man by presenting other, more sympathetic, versions: stories that stress the similarities between Mesmer and other doctors and that highlight political intrigues, social machinations and sexual discrimination. Note the plural, she reminds herself: no interpretations (not even her own) are definitively right.

And so she begins by writing a brief outline of the traditional Mesmer myth, the one in which he features as a disreputable but highly successful quack ...

* * * * *

By 1790, Franz Anton Mesmer (1734–1815) was notorious all over Europe. When Wolfgang Mozart wanted to poke some gentle fun at his fellow Freemason, he parodied the magnetic therapist in his opera *Così fan tutte*. Bumbling Dr Bezmer appears on stage wielding a giant magnet, recruited to cure two love-sick young couples who have developed unfortunate passions for the wrong partners. When this opera was first performed, the real Mesmer was lying low, trying to avoid the limelight after several years as Paris's stellar doctor.

Mesmer was a self-made physician from Vienna who arrived in Paris in 1778. He had already established his reputation as a successful magnetic therapist, and his first steps were smart ones: he opened a magnetic clinic in a fashionable quarter, and he took advantage of all the useful contacts he had managed to collect together. An expert publicist, Mesmer was soon attracting wealthy patients who had heard rumours about his cures for an impressive assortment of ailments, including dropsy, paralysis, gout, scurvy, blindness and deafness. Mesmerism, often called animal magnetism, became extremely fashionable – and Mesmer raked in the profits.

At this stage, Mesmer's most important equipment was his *baquet*, a large oval oak tub filled with magnets, iron filings, flasks of specially treated water, and aromatic herbs (Figure 23). According to his elaborate but ill-defined theories, a universal magnetic fluid circulated around the tub; patients who wanted more direct contact with the beneficial medium could, like the man on the left with crutches, tie themselves with ropes to the protruding iron bars. The wall mirrors reflected the fluid and intensified its effects, while gentle background music and aromatic scents perfuming the darkened room helped the healing processes along.

Mesmer also developed a more refined version of his therapy in which the *baquet* and other magnetic equipment was superfluous. Figure 24 shows a magnetic operator (as they were often called) who is using Mesmer's technique. No physical magnets are being used, but the male therapist gazes intently into the woman's eyes as

Figure 23: Franz Mesmer's magnetic clinic. Undated engraving by H. Thiriat. (Wellcome Library, London)

Figure 24: A mesmeric treatment. Ebenezer Sibly, *A Key to Physic, and the Occult Sciences*, London, 1794, p. 260. (Wellcome Library, London)

magnetic effluvia circulate between them under the control of his gesturing hands. Most of Mesmer's clients were wealthy women, and they were susceptible to so-called crises. This is the fate of the damsel in distress on the left of Figure 23, who is being cared for by Mesmer's assistants. These crises involved passing into a trance-like state, sometimes accompanied by convulsions, and they occurred after Mesmer had been passing his hands around his patient's body to realign the flow of magnetic fluid. Many women reported finding their crisis an extremely pleasurable experience which was both invigorating and relaxing; unsurprisingly, they came back for more.

Mesmer tried to have his treatments officially validated, but he repeatedly antagonised scientific inspectors with his slippery behaviour. Accusations of malpractice abounded: for instance, Mesmer was said to spike potions with purgatives, to pretend that magnetised water was having dramatic effects, and to be guilty of sexual impropriety (those crises were very suspicious). Nevertheless, backed by some wealthy financiers, Mesmer's practice boomed and his bank balance soared.

But everything suddenly changed in 1784, when a royal committee headed by Benjamin Franklin was appointed to investigate Mesmer's claims. They carried out a thorough round of tests, and concluded that magnets had no curative effects – any improvements were due merely to his devotees' overheated imaginations. After the team's damning report, Mesmer's meteoric career came to an abrupt end, and he fled from Paris. In the words of Thomas Jefferson, then representing America in Paris: 'Animal magnetism dead, ridiculed.'[1]

· CHAPTER 9 ·
THE RISE AND RISE OF
FRANZ ANTON MESMER

In vain do we boast of the enlightened eighteenth century, and conceitedly talk as if human reason had not a manacle left about her ... Mesmer has got an hundred thousand pounds by animal magnetism in France. Mainaduc is getting as much in London.

Letter from Hannah More to Horace Walpole,
September 1788

So many tracts and pamphlets were produced in the publicity battle between Mesmer's opponents and supporters that they now fill fourteen large volumes in Paris's Bibliothèque Nationale. Many of them are extremely abusive, but what should historians make of such prolific virulence? One answer is that this outspoken animosity indicates a vigilant scientific establishment successfully warding off a profiteering impostor. But in that case, it does seem rather strange that Mesmer's critics needed to expend so much energy on exposing an obvious fraud. Surely patients would soon realise that they were wasting their money?

A different answer is that other doctors reacted so strongly because they were frightened. To their alarm, Mesmer's treatments did seem to work; he was also creaming off some of their wealthiest patients. Still worse, splinter groups mushroomed all over France, and magnetic medicine became linked with radical politics. Castigating Mesmer as a quack was one way to get rid of him.

Mesmer appeared dangerous not because he was different, but because he was too similar. Some respectable doctors also endorsed using magnets, and Mesmer's nebulous magnetic fluid resembled many other invisible aethers supposedly circulating round the cosmos. Nowadays mesmerism would be classed as 'alternative medicine', but 200 years ago this term did not exist. There was no either/or situation: even marketing useless nostrums was not an infallible sign of quackery, since eminent physicians also bottled up their patent tonics for sale. Two famous examples are Sir Hans Sloane, President of the Royal Society, and Bishop George Berkeley, now better known as a philosopher.

Instead, it's more helpful to imagine physicians – like magnetic performers – as being laid out along a continuous spectrum of qualifications. At one end lay the society doctors who had been to university, belonged to professional associations and charged high fees; at the other were ill-educated men and women who were trying to scrape a living by caring for the poor or fleecing the rich. In between, all sorts of practitioners catered for different illnesses and budgets – surgeons, apothecaries, herbalists, midwives.

Enlightenment medicine was a highly competitive business. Although medical marketing remained unmonitored in Britain, in France the state authorities were trying to regulate it – hence the royal investigative committee that outlawed Mesmer. They wanted to set up firm boundaries between orthodox and pseudo science, between establishment and quack medicine, between lecturers and entertainers. Mesmer challenged their efforts by transgressing all these distinctions. Most threatening of all, his therapeutic powers set in question the supremacy of reason. Mesmer seemed to affect his patients' physical health by influencing their imaginations. But if this were possible, then animal magnetism would undermine rationality itself, since scientific detachment depended on the control of the body by the mind. That the arrow of influence should point the other way round was almost too scary to contemplate.

* * * * *

Although Mesmer had not been born into a wealthy family, his father used his connections well. He was gamekeeper to the Catholic Archbishop of Constance, who ensured that Mesmer received a solid Jesuit education in Germany and Bavaria. Mesmer was most interested in philosophy and theology, and he went first to the University of Ingolstadt (whose other notorious student was Victor Frankenstein). It was probably there that Mesmer first became obsessed with mysticism. This was a time when many young men – and a few women – were becoming fascinated by the occult. Underground societies of illuminati dedicated themselves to uncovering and tapping nature's mysterious hidden powers.

Far from being an uneducated charlatan, Mesmer studied medicine at Vienna, one of Europe's leading universities. When he graduated in 1766, he was a fully qualified physician whose dissertation was based on the work of Newton's own doctor, Richard Mead. His examiners were impressed by Mesmer's concept of animal gravity, which he introduced to explain how the stars affect people's behaviour. He argued that a universal harmonising power fine-tunes people's bodies as if they were musical instruments resonating in sympathy with the cosmos. Even though this now sounds outlandish, Mesmer's ideas were far more physically based than Mead's discussion of astrological influences. Mesmer's animal gravity was no vaguer than other Newtonian subtle aethers, and Newton himself valued universal harmony so highly that he gave the rainbow seven colours to correspond with the notes of the musical scale.

Two years later, Mesmer made a sensible move for a social climber: he married a rich widow ten years older than himself. He acquired a splendid home, furnished with its own well-equipped laboratory, and moved in aristocratic circles – the essential prerequisites for a flourishing medical practice. The Mesmers' mansion near the Danube attracted all the right people (including the musicians Joseph Haydn and Christoph Gluck), and the fashionable couple hosted many dinners and musical evenings. When the Mozart family visited Vienna for five months, Mesmer commissioned Wolfgang – then twelve years old – to compose one of his first operatic works and perform it to entertain Mesmer's guests. The game-keeper's son from Lake Constance had, it seemed, arrived.

But Mesmer was still ambitious. Like other doctors, he could often do little more than help his patients die comfortably. Many people were afflicted by chronic illnesses that could be neither diagnosed nor treated effectively, and rich invalids trailed from physician to physician, hoping to buy relief. Young women seemed to be especially prone to a panoply of transient but troublesome symptoms, including vomiting, convulsions, fainting and paralysis. When traditional prescriptions failed to alleviate their complaints, Mesmer searched for other approaches.

Mesmer learnt about a new therapy devised by Maximilian Hell, a Jesuit priest and professor of astronomy at Vienna University. Hell claimed that he had used magnets to cure stomach cramps, and Mesmer asked him to make some magnets for treating his own patients. Persuading one woman to act as guinea pig, Mesmer experimented by placing magnets on different parts of her body. At first she found this treatment very painful, but Mesmer decided that her violent reaction represented an essential breakthrough on the path to recovery. With his patient's cooperation, Mesmer gradually worked out how to use the magnets to achieve the most beneficial effects.

This might seem weird, but the basic idea of using magnets therapeutically was far from new. Look back to Kircher's frontispiece from the previous century (Figure 2, page 9). Written on the ribbon threaded through the clouds is a Latin motto meaning 'The world is bound with secret knots'. So although Mesmer declared that his theories were original, they did resemble older ideas that some sort of magnetic sympathy extends through the universe to unite people and animals with each other as well as with inanimate objects.

Magnetic medicine had a long history dating back to Egyptian times, when loadstones were said to draw out pain in the same way as they attract iron towards them. During the 17th century, the cosmic sympathy depicted by Kircher was believed to account for the beneficial effects of long-distance weapon salves: these were special (and expensive) sympathetic powders which were spread on a sword to ensure that the wound it had created would heal quickly. When a Dutch boy accidentally swallowed a knife, surgeons used a magnetic plaster to bring it nearer the surface before removing it from his stomach. This was probably the source of an English satire from 1720 on the Royal College of Physicians, who (supposedly) decided in a similar case 'to apply a *Loadstone* to his *Arse*, and so draw it out by a *Magnetick* Attraction'.[2]

Decades later, when Mesmer was at the peak of his fame, magnets were still being advertised to heal toothache, gout and rheumatism; women strapped magnets to their thighs to reduce the pangs of childbirth and avoid having a miscarriage. These home treatments were so widespread that Cavallo felt he needed to mock 'people who believe, that the application of the magnet cures the tooth-ach, eases the pains of parturient women, disperses white swellings, &c; and, on the contrary, that the wounds made with a knife, or other steel instrument, which has previously been rubbed with a magnet, are mortal'.[3] Aware of these traditional remedies, Mesmer continued to place his magnetic tub in the centre of his treatment rooms long after he had decided that physical magnets were unnecessary.

Mesmer was still in Vienna when he apparently discovered that he himself possessed a special type of magnetic power enabling him to magnetise everything he touched – paper, water, wood, animals. In particular, he claimed that his magnetic touch would cure the sick. His special ability was, he explained, rather like seeing colour – impossible to describe to a blind person but obvious to those who can experience it.

Mesmer's cures sound like the ancient laying on of hands, but he attributed his abilities to physical rather than mystical causes. He explained that he was redirecting the flow of universal magnetic fluid through his patient's body, so that it would circulate smoothly and restore the sufferer's natural state of harmony with the universe. Drawing an analogy with mineral magnets attracting iron, Mesmer described himself as a powerful animal magnet who could affect human beings, especially those who were ill. Just as ordinary iron rusts and loses its magnetism, he wrote, so 'the universal fluid, destroyed or weakened in a sick body, must be strengthened with additional fluid in order that the body be able to regain its former vigour and have the obstacles removed'.[4]

Despite the ancient roots of his ideas, Mesmer aligned himself with contemporary science and progress. He complained about old-fashioned reactionaries who refused to contemplate his comprehensive system which was, he insisted, based solidly on the principles of physics. 'I will present a theory ... which is as simple as it is new', he later proclaimed; 'I will substitute for the vague principles which, up till now, have served to guide medicine, a method which is simple, general, and represented in Nature.'[5] His animal magnetism sounds as if Kircher, Descartes and Newton have been blended together: Mesmer described a subtle fluid which mediates between the stars, the earth and living creatures, which acts at a distance and insinuates itself into the human nervous system so that, like animate loadstones, people have opposite poles.

Epilepsy, haemorrhoids, menstrual disorders, convulsions, fevers – Mesmer's list of successful cures was impressive. Although it seems sensible to be suspicious of his own case histories, Mesmer did manage to collect some stunning affidavits. One teenage girl in the emperor's household had been struck blind at the age of three. Physicians had tried every available treatment – leeches, cauterisation, purgatives, electricity – but these probably made the situation worse: by the time that Mesmer saw this gifted young pianist, her eyes were grotesquely distorted. At first Mesmer's magnets produced some alarming and painful symptoms, but within a few weeks his patient could see again. Mesmer embarked on a programme to retrain her body and her mind to cope with this transformation; she was, he promised, well on the way to total recovery.

At least, that was her father's first account of what happened. But when other doctors made scathing comments about Mesmer's intentions as well as his healing powers, the man changed his mind and brought his daughter home. Desperate to escape the scandal, Mesmer set off for Paris with his magnetic equipment.

* * * * *

Good contacts always help. Equipped with a letter of recommendation to the Austrian ambassador, Mesmer set up practice in a prestigious quarter of Paris, and the clients started pouring in. Within a few months, he expanded to larger premises so that he could give private consultations for rich customers in addition to the cheaper communal sessions round his magnetic tub.

For individual treatments, Mesmer sat facing his client, knees touching, both of them lined up north–south, and staring intently into each other's eyes. As he passed his hands around his patient's body, he paid particular attention to afflicted areas. Although some people remained unaffected, many of them reported feeling changes in temperature as well as the crisis which was induced (according to Mesmer) by a surcharge of magnetic fluid. These health-giving crises sometimes turned into convulsions, and in the heady atmosphere of the public salons they could spread like a contagion.

Although Mesmer never managed to speak good French, he soon learnt about the psychology of rich Parisians. To advertise his philanthropic credentials, Mesmer reserved one of his four tubs for poor people and made no charge. Few customers took advantage of this democratic generosity, preferring instead to pay large fees for gathering round the other three tubs; the need to reserve places well in advance boosted the desirability of being seen in the right treatment room. The German doorman announced visitors with a whistle – three different pitches to warn Mesmer of the arrival's social status. Only the most distinguished ladies were conducted to the special *baquet* that had been decked with flowers.

A visit to Mesmer's clinic could be literally a life-changing experience:

M. Mesmer's house is like a divine temple upon which the social orders converge: abbés, marquises, humble seamstresses, soldiers, doctors, young girls, midwives, the dying as well as the strong and the vigorous – all drawn by an unknown power. There are magnetizing bars, closed tubs, wands, ropes, flowering shrubs, and musical instruments including the harmonica, whose piping excites laughter, tears, and transports of joy. Add to these objects the allegorical paintings, padded consulting rooms, special places designated for crises, a confused mixture of cries, hiccups, sighs, songs, shudders ... one finds at M. Mesmer's only creatures given over to pleasure or to hope.[6]

This tongue-in-cheek description (the verbal equivalent of Figure 23) is a gentle example of the countless satires that appeared – poems, plays, caricatures, spoof manuals. So if Mesmer was so easy to satirise, why was he so successful?

One answer is that magnets were already being used medically. They had been introduced in England to cure toothache about ten years earlier, and after a shopkeeper was spotted holding a magnet in his mouth, the Parisian Royal Society of Medicine commissioned two doctors to investigate. Their report was 150 pages long, and it unequivocally endorsed the use of magnets for treatments. These medical inspectors compiled evidence from all over Europe, and they commented favourably on Hell as well as Mesmer (who had helpfully sent over some of his magnets – why waste a good advertising opportunity?).

Case-study after case-study made it clear that the calming effect of magnets could be enormously beneficial, especially for spasms, convulsions and sharp pains. Being French, the two physicians gave a Cartesian-style explanation: the magnetic fluid acts directly on the nerves, and so could relieve complaints caused by nervous diseases. Since the symptoms returned when the magnet was removed, they recommended strapping magnets closely to afflicted parts of the body.

Different styles of therapeutic magnets were available to treat different illnesses. In Figure 25, taken from the official medical report, designs 1 to 3 are made from small bars covered in black velvet, and they could be worn as necklaces or headbands. The shaped plaques 4 to 6 were carried next to the skin for a stronger effect on specific areas such as the arm, heart or thigh: the punched holes were for ribbons. Other models included pointed bars for troublesome teeth, curved magnets to loop round deaf ears, and a strong battery of eight magnets to alleviate headaches.

Like James Graham's Celestial Bed, Franz Mesmer's magnetic tub was not self-evidently ridiculous. The Parisian report enthused that future research would forge ahead because Gowin Knight's artificial magnets had introduced new possibilities. Could cures perhaps be intensified by sleeping in magnetic beds or sitting on magnetic chairs? These were not the utopian dreams of mercenary quacks, but the realistic projects outlined by representatives of France's major medical society.

Another reason for Mesmer's popularity is that he arrived in Paris at the right time. Many patients were disillusioned by the harsh drugs and painful techniques administered by other doctors at great expense but with little obvious benefit. Instead, they were searching for less invasive cures that would restore their harmony with nature through a healing process based on a close relationship with the therapist. Mesmer lavished attention

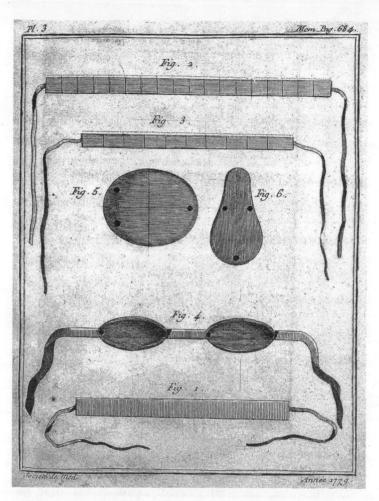

Figure 25: Therapeutic magnets. C. Andry and M. Thouret, *Observations et Recherches sur l'Usage de l'Aimant en Médecine*, 1782, p. 684, Plate 3. (Wellcome Library, London)

on his clients, and prescribed few medicines or diets. Being roped to a *baquet* was far less unpleasant than receiving a massive shock from an electrical machine or swallowing a punitive dose of purgative potion.

Most importantly of all, Mesmer attracted patients because his treatments worked – rich people do not pay out small fortunes to therapists for several years unless they feel better, even if only temporarily. He provided testimonial after testimonial from grateful patients, such as this one from a 40-year-old military officer who, after weeks of undiagnosed illness, was dismayed to find himself staggering around like a drunkard: 'without treating me with drugs or any other remedy than what he calls animal magnetism, he caused me to feel powerful sensations from head to foot ... Four months later, I can stand erect and stable on my feet. My head is steady. I have regained the use of my tongue and can speak as intelligibly as anybody ... In short, I am entirely free from all my infirmities.'[7]

All but the most hardened critics had to admit that many of Mesmer's patients thrived under his care. Sometimes chronic symptoms disappeared after years of suffering; other sick people were less fortunate, but they welcomed any brief respite that they could find. Modern-day patients expect their doctors to cure them. During the Enlightenment, when there were neither antibiotics nor anaesthetics, expectations were lower. It often seemed worth paying for a restorative magnetic treatment that did bring some relief – whatever the reason.

* * * * *

Being rich and famous was not enough for Mesmer: he also craved official recognition. His first important convert was Charles d'Eslon, a doctor at the University of Paris who became convinced that animal magnetism worked. He described how Mesmer had healed one of his own patients, a young boy who was obviously dying from a long bout of stomach upsets and fever. A few hours after being infused with some of Mesmer's powerful personal magnetism, the child was dining off crab and champagne; a month later, he was completely back to normal.

Despite such spectacular recoveries, the sceptics remained unconvinced. Mesmer tried hard to win them over and as controversy escalated, he won one important round against his critics by threatening to leave the country. Only after some audacious bargaining with Queen Marie-Antoinette (well publicised by Mesmer, of course) did he eventually consent to stay. But a year later, Mesmer made a tactical error – he left Paris for a few weeks, and D'Eslon promptly set up his own rival practice. Mesmer was rescued by the Society of Universal Harmony, founded by two wealthy businessmen who made him an offer he reluctantly accepted – in exchange for financial security, he would divulge his secrets to the Society's members.

D'Eslon was a skilled, dedicated doctor who enjoyed making money but who also genuinely wanted animal magnetism to be recognised so that more people could benefit from its healing powers. Altruistic and confident of his own abilities as a magnetiser, D'Eslon asked the government to set up an investigative committee. When these state inspectors started work in March 1784, Mesmer refused to have anything to do with them. He probably guessed that disaster lay ahead – and five months later, it arrived.

· CHAPTER 10 ·
ANIMAL MAGNETISM EXPOSED

And now the solemn farce began:
Slow from his seat arose the wonder-working man …
Now he approach'd the 'subject to be treated,'
The 'dauntless fair' alone unmov'd appear'd,
And boldly dar'd the awful test;
And now his dexter hand the Doctor rear'd
Now darts it down, like lightning, at her breast.
Laurence Hynes Halloran, *Animal Magnetism.*
The Pseudo-Philosopher Baffled, or, The Biter Bit.
A Comic Morning Adventure, 1791

The royal report condemning mesmerism was published in August 1784, and the satires were savage. One caricature shows Mesmer, laden down with money, flying away on a broomstick and a donkey. Below him lies a shattered *baquet*, crushed underneath a voluptuous, blindfolded travesty of Truth, the goddess who conventionally radiates illumination. Instead, the dazzling light of reason shines forth from the report brandished triumphantly by Benjamin Franklin, the investigative committee's official head. This sketch's message is clear: crumbling under the glare of Enlightenment rationality, the dark forces of superstition and magic will be defeated.

Franklin was then almost 80 years old and crippled by gout, so the effective leader of the nine-man team was the chemist Antoine Lavoisier. The other members included the astronomer Jean Bailly and also Joseph Guillotin, selected because he was an eminent doctor, although now he is more famous as a very special inventor – only ten years later, Bailly and Lavoisier were both guillotined during the Terror of the French Revolution.

In pamphlets and newspapers, Mesmer was caricatured as a diabolic monster or a donkey, traditional symbol of a charlatan. Even enlightened men of reason seem to have enjoyed pornographic humour, and Figure 26 shows a donkey-like Mesmer with an exaggerated forefinger looming lewdly over a semi-conscious young woman close to a crisis. Transfixed by his gaze, she has clearly succumbed to his power, and the dream-like atmosphere is enhanced by mythical creatures lolling in the clouds. As she reclines seductively in an armchair, the apparently casual position of her right hand emphasises the sexually charged nature of allegations against Mesmer.

Beneath the caricature's title – *The magic finger* – are the Latin words *Simius semper simius. Semper* means 'always', and *simius* means both an imitator and an ape. This is a characteristic Enlightenment pun, mocking Mesmer as a con-man as well as a monkey. Depicting magnetisers as animals was far more shocking then than now, because the evolutionary ideas introduced by Charles Darwin still lay far ahead in the future. At the end of the 18th century, most Europeans believed that God had created human beings and animals separately; to suggest any family relationship with apes would have been unthinkable.

It's hard to be sure exactly what went on inside mesmeric salons, because most of the descriptions are biased either for or against animal magnetism. But the following report – taken from an unpublished handwritten letter about an English magnetiser – does sound like a reasonably

LE DOÏGT MAGÍQUE
OU LE MAGNÉTISME ANÍMAL
Simius semper Simius

Figure 26: The magic finger or animal magnetism.

honest account of how, after conventional treatments had all failed, one sick woman eventually found relief through experiencing a crisis:

> But the 2ᵈ day of her Attendance with all her resolution to the contrary she was totally conquered … When only with his Hands he directs the Fluid to her, after about a Quarter of an Hour she begins to feel uneasy, as if compressed by something & by increasing weight over & round her Head, until at last unable to support the burthen down she sinks into a dose, in sometime she awakes from this Slumber, in great perturbation even to Hystericks, & convulsions … from all this distress & Misery, she feels intirely Calm; and is comforted by a Glowing Warmth from Head to foot, which she sais is most reviving.[8]

In spite of such testimonials, all the caricatures, skits and pulpit denunciations vibrate with emotional energy. Many Parisians were outraged by Mesmer's activities, but why should the public disgrace of a medical practitioner arouse such fascination? At the same time, why did so many critics denounce the committee's conclusions? The report itself, which describes the committee's tactics, reveals some of the answers.

* * * * *

To start with, it seems strange that this committee – and also a smaller medical one appointed later – readily accepted that there would be no access to Mesmer himself, who refused to cooperate. The matter was deemed so urgent that they were told to work with

D'Eslon instead. It was obvious that after the reports were published, Mesmer could simply retort that his version of animal magnetism was different from that of his rival.

Choosing to ignore that problem, the team set to work. Equipped with compass needles, electrometer and thermometers, they explored the secret contents of D'Eslon's *baquet* and observed the enthusiasts clustered around it. Regarding these paying customers as though they were experimental objects, the commissioners were amazed by the crises they witnessed, especially among the women. Some patients were soothed into repose, while others became extremely agitated, convulsing for hours on end and sometimes coughing up blood. Undeniably, something was happening. But what? And how was it possible to prove an apparently unprovable negative – that an invisible, weightless fluid did not exist?

Instead of exploring why mesmeric treatment worked or how it could be made useful, the investigators chose to focus on pinning down the ineffable magnetic fluid. They decided to experience for themselves what it felt like to be magnetised, and they arranged for D'Eslon to give them private sessions. So instead of relying on scientific apparatus, they had to turn their own bodies into instruments and observe themselves with clinical detachment. This project immediately raised problems of verification and calibration: how can you be sure that one person's pain is worse than his colleague's? Although their report is couched in formal language, it's obvious that some interesting discussions arose as they focused their attention inside their own bodies. Was X's twinging gut and aching head the result of overeating, or had the magnetic influence been at work? Or did he just moan more loudly than Y? After all, they noted, one of the team (discreetly left anonymous) was often subject to attacks of that conveniently vague complaint, nervous irritation.

In these pre-Revolutionary years, aristocratic hierarchies still ruled. God had created humans and animals separately, but He had definitely made some people better than others: men were superior to women, and the nobility had been born to lord it over their servants. The investigators concluded that their own robust, elite bodies were impervious to the invisible fluid's power, so they recruited seven lower-class invalids in order to find out whether such inferior, less rational specimens would be more susceptible. When three of them did report feeling effects, the team selected some more trustworthy volunteers 'from the polite world who could not be suspected of sinister views, and whose understanding made them capable of inquiring into and giving a faithful account of their sensations'.[9]

Eventually five out of fourteen sick people experienced magnetic sensations, but their testimony was dismissed: a hot knee was deemed to be trivial, the three working-class witnesses were said to be obviously unreliable (so what was the point, one wonders, of summoning them in the first place?), and as for Madame de V. with a nervous disorder who became dejected – well, it was hardly surprising that a woman with those ubiquitous 'irritable nerves' would not be able to sit still for over an hour without feeling rather strange. In contrast, the commissioners who had 'armed themselves with that philosophic doubt which ought always to accompany inquiry, have felt none of those sensations'.[10]

Blind testing – trickery – was the next stage in the

programme: truth was to be reached through deception. Gratifyingly for sceptical inquisitors, D'Eslon's prize student fell into a crisis under an ordinary tree that had not been magnetised in advance. One woman had convulsions when she thought she was being offered magnetised water, but calmly swallowed a cup that had been treated; and another female guinea pig felt heat in the wrong parts of her body when she was blindfolded to prevent her from knowing exactly where the magnetiser was making his passes.

The report's verdict was clear: although everyone is surrounded by a personal cloud of heat and sweat, animal magnetic fluid does not exist. The distressing symptoms that sometimes accompanied a crisis – diarrhoea, bloated intestines, spasmodic vomiting – were attributed to over-dosing on laxatives and the pressure of the magnetiser's hands. But the inspectors singled out one prime culprit for condemnation – the imagination. Although D'Eslon did point out how beneficial this therapeutic tool might be if physicians would only learn how to control it, the committee insisted on denigrating the imagination as an 'active and terrible power' that demanded constant vigilance to prevent it from overwhelming even those philosophical men who normally behaved rationally. If imagination were allowed to flourish, worried Franklin, then atheists would have a new weapon for attacking the Bible, because they might argue that the genuine miracles due to God had occurred only in the witnesses' minds.

Animal magnetism's supporters rushed into print to attack the committee's findings. Mesmer himself, as foreseen, dissociated himself from D'Eslon – but he also continued to attract a large clientele. Grateful patients produced long testimonies of the treatment's efficacy, while natural philosophers emphasised the similarities between mesmerism and contemporary physics. One eloquent lawyer advocated Mesmer's non-invasive treatments by attacking conventional science in the voice of Mother Nature: 'Remember that for four thousand years you have not ceased to disturb, torment, shatter all my works ... Have you healed any more ills than I have? Have you at least assuaged them? Have you not, on the contrary, augmented them?'[11] Other physicians agreed with this plea on Nature's behalf. By then, animal magnetism was widespread not only in Paris, but also in large provincial cities as well. D'Eslon spoke for many when he insisted that because the therapy worked, he would go on using it.

Well-informed critics accused the committee of distorting the evidence, and they posed some sharp questions. Why had the inspectors set up their own punitive conditions instead of ones favourable to the effects they were searching for? How could they draw such far-reaching conclusions on the basis of so little evidence? Why were they so eager to dismiss the cures? Much as the commissioners boasted about their cool objectivity, it seemed clear that they had pretty much made up their minds in advance. They had not even watched Mesmer in action, yet they were ready to outlaw his practices.

* * * * *

As the investigators' comments on class imply, the real issue at stake was power. Patients in the throes of a crisis were, they noted in horror, 'entirely under the government of the person who distributes the magnetic virtue'.[12] Any natural philosopher who yielded to the fluid's influence would lose control over himself; he would (awful thought) run the risk of behaving like a woman or a member of the working classes. By succumbing to the magnetiser's strength, a rational man would lose his very identity as an independent individual.

The ramifications were terrifying. Instead of undergoing the years of university training needed for a medical degree, magnetic therapists claimed to manipulate their clients through accessing a special property which they had possessed since birth. If this were true, then the country might become governed by poorly educated animal magnetisers rather than by the aristocratic elite – so satirists revelled in parodying footmen who insulted their employers and criminals who sentenced their magistrates. Lavoisier painted a grim scenario: if people learnt how to magnetise themselves, he wrote, then 'we would have to shut the schools, change the system of instruction, and destroy those bodies regarded till now as the depositories of medical knowledge'.[13]

To drive their message home, sceptics compared animal magnetisers to puppet-masters. Paris was the world's capital of marionettes and automata, and enlightened natural philosophers delighted in exposing the hidden machinery that made artificial ducks defecate or toy drummers drum. They decreed that such entertainments belonged to conjurors and street theatre. When Mesmer controlled the subtle magnetic fluid, he performed like a puppeteer pulling invisible strings; because he refused to be confined within the shady world of magical entertainers, he was transgressing social boundaries policed by elite physicians. Such comparisons later inspired the poet Samuel Taylor Coleridge to thunder during a lecture that Prime Minister William Pitt was 'the great political Animal Magnetist, who has most fully worked on the diseased fancy of Englishmen; and by idle shew, and alarming bustle, and many a mysterious trick has thrown the nation into a feverish slumber, and is now bringing it to a crisis which may convulse mortally!'[14]

The committee also had a hidden agenda. They wrote a secret report, but it was for the king's eyes only, and it remained unpublished until years later. This revealed how worried these men were about magnetic morality, and the power that a male magnetiser might exert over his female patients. During the commission's enquiries, a policeman asked D'Eslon whether women were in danger of being sexually abused. His answer was not very diplomatic – that's precisely why we have public salons, he explained. But what, the inspectors wondered, might be going on in the private treatment rooms?

The astronomer Bailly watched carefully (voyeuristically?) as women approached a magnetic crisis: 'the face reddens and the eyes become ardent – this is a sign from nature that desire is present ... the pupils become moist and breathing becomes shallow and uneven. Then convulsions occur ... This disorder is not perceived by the woman but is obvious to the medical observer.'[15] Since women were paying to repeat the experience, they probably perceived more than Bailly appreciated. Nevertheless, he maintained that it was his duty to protect innocent, sexually naive women who did not realise what was happening to them. Even the magnetiser himself was at risk when sitting in the intimate therapeutic position, knees and feet touching, since a sensitive, responsive woman might exert a reciprocal malign influence and invert the direction of power. In this authoritarian, hierarchical society, young women needed to be kept under control.

The commissioners slated imagination because they regarded it as a feminine attribute, one that accounted not only for women's susceptibility to a magnetiser's charismatic charms, but also for their flights of fancy, addiction to Gothic novels and liability to indefinable nervous illnesses. The imagination was said to be wired up to the eyes – hence the significance of the magnetiser's gaze – and also to the uterus, which had long been believed to exert an evil influence over the entire female frame.

Since women were held to be such irrational creatures, they could not, of course, be expected to look after themselves. One doctor was so appalled by the popularity of animal magnetism that he decided to expose and exploit the female gullibility he saw around him. To help resolve one woman's illness during pregnancy, he called in a healer he denigrated as a 'sorcerer'. 'My imbecilic patient was healed suddenly,' he reported scathingly, 'by the very fact of her stupidity.'[16] Along with many other men who claimed to be enlightened, this provincial physician condemned Mesmer as a trickster but blamed witless women for his success.

· CHAPTER 11 ·
MESMERIC MOVEMENTS

I continue to make use of the wonderful power that I owe to M. Mesmer, and I bless him for it every day, because it makes me useful and I am able to help and benefit many ill people in the neighbourhood ... I have only one regret – not being able to touch everyone.

Armand de Puységur, *Memoirs towards the History and Foundation of Animal Magnetism*, 1784

Three months after Franklin's report was published, a burlesque comedy called *The Modern Doctors* opened in Paris. It was an immediate hit. Night after night, audiences flocked to cheer at the downfall of the money-grabbing Cassandre (Mesmer) and his evil confederate (D'Eslon). One mesmeric devotee did try to sabotage the performances; unfortunately, the servant he sent along failed to realise that a double bill was showing, and he heckled at the wrong play.

But outside the theatre, the real-life Mesmer and D'Eslon were thriving. Although Mesmer himself left Paris the following summer, societies dedicated to animal magnetism were being set up all over France, and soon different variants were being practised throughout Europe.

The enthusiasts who adopted Mesmer's ideas also adapted them for their own purposes. As a result, splinter groups were founded whose beliefs diverged widely. Confusingly, people who called themselves mesmerists were not all doing the same thing; conversely, people who denied that they were mesmerists were actually engaged in activities closely resembling those of Mesmer himself. Some therapists were clearly in it for the money, while other practitioners belonged to radical political movements, or were secretly signed up in mystical and masonic associations. In very broad terms, versions of mesmerism continued to thrive in France and Germany, whereas in England, animal magnetism petered out after a brief flurry of interest but became fashionable once again from the 1830s.

For the sceptics who remained unconverted, 'mesmerist' and 'magnetiser' became smear terms that came in handy for abusing more or less anyone involved in something unusual. When John Robison, Scotland's paranoid magnetic expert, was searching for evidence of an international underground conspiracy, he lumped together 'the fanatical and knavish doctrines of the modern Rosycrucians – by Magicians – Magnetizers – Exorcists, &c'.[17] In England both political parties accused the other of being linked with animal magnetism. Figure 27 shows a Whig satire on the Tories. Skinny William Pitt limps in to one of London's best-known magnetic surgeries, begging not only to be cured of his gouty leg but also to be relieved from 'the Fumes of discontent' – a reference to his recent unpopular proposal to tax tobacco.

Propagandists were using animal magnetism to suit their own agendas. Similarly, modern historians have interpreted the activities of Mesmer and his followers to impart particular messages. Now that it's no longer fashionable to jeer at Mesmer as an out and out charlatan, scholars have chosen different ways to redeem his reputation. Occasionally their sense of mission seems to

Figure 27: Billy's Gouty Visit, or a Peep at Hammersmith. © British Museum

overwhelm their learned vision: it's tempting to make magnetisers appear everywhere. Thus according to one academic, the son of General Charles Rainsford (Governor of Gibraltar and cousin of the Royal Society's President) was an avid convert. A sense of humour might have helped this analysis. In fact, the young man's letter, which is now in the British Library, reported only that an animal magnetic conference with his girlfriend nine months earlier had recently resulted in its anticipated but unwelcome outcome.

With an approach reminiscent of Robison's, some analysts place Mesmer in the midst of a covert mystical tradition going back for centuries and embracing William Blake as well as William Yeats. In contrast, other experts establish Mesmer as the founding father of modern psychoanalysis, tracing out a direct line of influence through to Sigmund Freud. Other historians are more interested in demonstrating the interaction between scientific pronouncements and political aspirations, and so they have linked mesmerism with radicalism. To make the situation still more complicated, detailed studies within particular countries have revealed how risky it is to make any sweeping international generalisations.

In other words, there is no single right way of describing what happened next. But there are several interesting ways of thinking about it ...

* * * * *

In Britain, therapists aimed above all to make money. Because they arrived after news of Mesmer's disgrace had seeped across the Channel to London, only a few of these medical entrepreneurs claimed to follow his principles closely. The most important was the Reverend John Bell – aka Monsieur le Docteur Bell – who recreated a miniature Paris in Covent Garden, then rather a seedy area. Bell's sales pitch claimed that he had been trained in France, and he built his own oaken tub lined with loadstones and artificial magnets. To top up funds, he wrote and translated mesmeric literature, but after a few years he was squeezed out of business and probably emigrated to America.

Bell's most famous rival, John de Mainauduc, succeeded because he chose very different tactics. As one measure of his reputation, a couple of hundred people paid in advance for his book on animal magnetism, which was published posthumously. Significantly, De Mainauduc explicitly and repeatedly distanced himself from Mesmer and magnetic apparatus. Helped by a female assistant (probably his live-in partner), De Mainauduc ran a sober establishment in which he treated his patients by gazing into their eyes and using his hands to redirect effluvial particles through their bodies, as in Figure 24, page 150 (which is an English picture).

De Mainauduc was probably an Irish draper's son, although he also spread rumours of more exotic origins. After a conventional training as a surgeon at leading London hospitals, he studied medicine in France, where he came across D'Eslon and recognised the earning potential of animal magnetism. Although his rush back to England was delayed by a spell of calm weather at Calais, De Mainauduc passed the time by picking up some impressive testimonials of his magnetic cures, which he promptly published (embroidered?) as soon as he got back to London. Installing himself in fashionable

Bloomsbury Square, De Mainauduc set up shop and watched the customers arrive.

London's Royal College of Surgeons still owns De Mainauduc's handwritten list of patients and students. Many of his faithful clientele were rich but chronically ill, and included leisured members of the aristocracy as well as newly wealthy wives of Quaker industrialists. In addition, rivals came to learn his techniques and start their own practices, although none of them were as successful. De Mainauduc also attracted a small artistic group fascinated by the occult. To satisfy their cabalistic interests, he suffused his lectures with an esoteric aura, and many members of international masonic cults entered his magnetic treatment rooms.

The most notorious of De Mainauduc's protégés was Philip de Loutherbourg, a theatrical designer and artist from Alsace. His mystical friends included the Cosways, both eminent painters and both keen animal magnetisers. Maria Cosway's enthusiasm embarrassed Cavallo, who belonged to the same émigré circle, and presumably also dismayed her transatlantic lover Thomas Jefferson. Although he had confidently pronounced French mesmerism to be dead, she sent him a guilt-tripping moan that in the absence of his long-awaited letter, her magnetic susceptibility to London's depressing influence was making her miserable.

De Loutherbourg rashly promised to provide treatments free of charge at his Hammersmith house, but the neighbours soon complained about the crowds and called in the police. In Figure 27, De Loutherbourg stands in front of his grotesque working-class patients and addresses Pitt in Germanic English; the scroll advertises

'CURES by a TOUCH' (a multiple pun on magnetising iron, fraud, madness and sexual misbehaviour) for ailments including 'a nine months Dropsy', lethargy in bishops and blindness in statesmen.

De Mainauduc's prescriptions resonated closely with those of other doctors. Like them, he stressed the need to augment Nature's own healing powers by restoring a state of equilibrium and removing obstacles to the healthy circulation of particles. His theories mirrored the Quaker approach to commerce: they preached that allowing the free activities of self-motivated traders would contribute to the well-being of society at large. As the king wandered in and out of sanity, Londoners welcomed De Mainauduc's reassuringly physical explanations of illness in terms of atomic chains. His insistence on transfixing his patient's eye (as in Figure 24) was also familiar: the provincial physician Francis Willis had temporarily cured his royal patient by subjecting him to his own stony gaze and forcing him to succumb to his commoner's will.

Satirists made plenty of money from animal magnetism. The playwright Elizabeth Inchbald translated a French farce about Mesmer and passed it off as her own; years later, it was still playing in a double bill with a farewell performance of Sarah Siddons starring in *Macbeth*. Gossip columnists sniped at fashionable figures thinly disguised as Major MacNeedle and Lady Bumbustle enjoying magnetic conferences ('conversation' was an old euphemism for sexual rather than verbal intercourse). The poet Robert Southey condemned De Mainauduc at length in his famous catalogue castigating English eccentricities, cleverly written as though he were a naive Spanish traveller.

Partly because of such denunciations, animal magnetism started to enter everyday language. Southey's friend Coleridge had been a London schoolboy at the peak of De Mainauduc's success, and his early description of how the ancient mariner subdued a wedding guest with his glittering gaze resembles De Mainauduc's control over his patients:

For that which comes out of thine *eye* doth make
My body and soul to be *still* ...[18]

The new vocabulary was even adopted by the translator of Franklin's damning report, William Godwin (Mary Shelley's father). Godwin had written an introduction for his British readers that celebrated Mesmer's disgrace as the triumph of Enlightenment rationalism over superstition. Nevertheless, he later wrote a letter reassuring his wife that although they were physically separated, animal magnetism was still drawing them together.

Animal magnetism was savaged because it came from France. This was seen as the land of political as well as scientific revolution, and defenders of British values – chauvinists like John Robison – were determined to keep out any ideas which might alter the status quo. Animal magnetism became closely linked with two other continental innovations: Lavoisier's new French chemistry, which had eliminated phlogiston; and galvanism, the Italian discovery of electricity in living bodies. Chemistry and galvanism later became part of mainstream science, but during the 1790s they were denigrated by being linked with quackery and revolutionary politics.

Opinions about science often depend on political allegiances. Galvanism, mesmerism and anti-phlogistic chemistry were – like algebra – regarded with suspicion because they had been imported from the continent and differed from traditional British science. Associated with radical beliefs, they opened up the frightening possibility that a new hierarchy might be established, one that was based on being born with intellectual ability rather than inheriting landed estates.

A prime target for reactionary men of science was Humphry Davy, who taught chemistry at London's newly founded Royal Institution, but was also notorious for his association with a radical medical group sympathetic to the French Revolution. The Royal Institution was originally intended to be a charitable organisation devoted to promoting popular education about science, such as showing how chemistry could improve agricultural production, but it soon became a fashionable lecture centre for the middle classes. Davy later became President of the Royal Society, but as the new century opened he represented all that was threatening in science. As well as teaching French chemistry, he forecast that electricity would provide the key to unlock Nature's deepest secrets – Davy is often cited as Mary Shelley's model for Dr Frankenstein.

The anonymous author 'Sceptic' aimed his fantasy about 'The Birth of Wonders!' directly at Davy. His satirical fable opens with the arrival of the French Revolution, followed by its little sister '*Mesmeria* ... her eye fascinated and charmed to the spot ... the mysterious power of inchantment, subverted the laws of nature.' The next baby wonder leaps into a frog Luigi Galvani is eating for dinner, while young Antiphlogiston sets about duping all the gullible men of science.[19]

The Royal Institution itself also came under attack. In a cruel caricature (Figure 28), conservative James Gillray savaged a demonstration by Davy and Thomas Young, the Royal Institution lecturer who appeared earlier in *Fatal Attraction* as the originator of Figure 9 (page 78) and a magnetic expert (Young was one of those depressing people who are knowledgeable about practically everything). As Young administers nitrous oxide (laughing gas) to an exploding volunteer, Davy wields a large pair of bellows. The man with a bulbous nose at the back right is the inventor Count Rumford, who married Lavoisier's widow and waltzed off with a large financial settlement

Figure 28: *Scientific Researches! – New Discoveries in Pneumaticks!* Hand-coloured etching by James Gillray, 1802.
© British Museum

after their divorce a couple of years later. Like other financial supporters of the Royal Institution, Rumford benefited by advertising his philanthropic credentials.

Critics disliked the Royal Institution because it invited non-experts to learn about the latest scientific research: in its earliest days, even artisans were permitted to enter the galleries. To satirise the audience, Gillray has exaggerated physiognomies and costumes. Southey divided the spectators into two groups – bored men and superficial women. 'Part of the men were taking snuff to keep their eyes open', he reported; 'others more honestly asleep, while ladies were all upon the watch, and some score of them had their tablets and pencils, busily noting down what they heard, as topics for the next conversation party.'[20] Southey drew little distinction between credulous fans of animal magnetism and fashionable visitors to the Royal Institution.

* * * * *

The fate of animal magnetism was very different in France. Mesmer himself fled the country the summer after Franklin's report, but variations of his ideas flourished in the provinces. One schism arose from Mesmer's own greed. He soon parted company with his backers, the Society of Harmony, because he wanted to keep exclusive control over his theories; in contrast, his financial rescuers felt that by bailing Mesmer out of trouble, they had earned the right to make his techniques public. Inspired by egalitarian ideals, they wanted to make this marvellous new therapy available to all: they felt it was wrong to charge such high prices for a treatment that promised to transform the nation's health.

The Parisian Society of Harmony split apart when a radical faction was excommunicated by Mesmer's wealthy aristocratic clients. They rapidly lost interest in the details of mesmeric therapy, and focused instead on fomenting political unrest. For them, animal magnetism promised an egalitarian type of science, a way of enabling everybody to share in rational explanations for nature's invisible secrets. Transforming society, they maintained, went together with a change in scientific outlook – and Mesmer's magnetic medicine promised a democratic form of treatment that would challenge the stranglehold exerted by society physicians and their rich clients.

One of the most prominent members of this splinter group was the Society's co-founder Nicolas Bergasse, a wealthy lawyer who had defended Mesmer against D'Eslon by initiating twelve disciples (a significant number) for a high fee. Bergasse regarded himself as the mesmeric counterpart of Jean-Jacques Rousseau, the reforming political philosopher whose slogan 'Liberty, Equality and Fraternity' became the rallying call of the French Revolution. Bergasse effectively recreated Kircher's interconnected universe (Figure 2, page 9) by explaining that people and planets influence each other through the motion of a cosmic magnetic fluid.

In predicting that France could regain its former state of primitive harmony, Bergasse echoed Rousseau's emphasis on education. He explained that surrounding a child with the right carers meant that its small body would be suffused with beneficial fluid flowing freely through it. For Bergasse, moral and physical improvement were magnetically tied together, so that sick patients or sinful criminals could also be improved by constant subjection to powerful fluid influences. Ultimately, he believed, purifying the nation's health would result in a behavioural revolution, so that France's political institutions would be transformed.

Whereas Franklin and Lavoisier viewed Mesmer as an upstart outsider who should be suppressed, it was precisely Mesmer's anti-establishment stance that attracted young activists. Consider Jacques-Pierre Brissot de Warville. Despite his impressive name, Brissot was the thirteenth child of a provincial bar-owner, and he adopted the aristocratic suffix (de Warville) in his vain attempt to enter Parisian society. After joining forces with Bergasse, Brissot played a leading role in the Revolution as a prominent Jacobin. 'The domain of the sciences must be free from despots, aristocrats and electors', Brissot proclaimed. Like Bergasse, he was swept away by the ideological appeal of Rousseau as well as of Mesmer: 'But I, a father who fears doctors, I love mesmerism because it identifies me with my children. How sweet it is to me ... when I see them obey my inner voice, bend over, fall into my arms and enjoy sleep! The state of a nursing mother is a state of perpetual mesmerism. We unfortunate fathers, caught up in our business affairs, we are practically nothing to our children. By mesmerism, we become fathers once again. Hence a new benefit for society, and it has such need of one!'[21]

Across the Channel, animal magnetism mainly remained based centrally in London; in contrast, French mesmerism extended far out of the metropolis. In 1785, the Royal Society of Medicine was aghast to discover that all the major provincial cities had their own Societies of Harmony. With fewer aristocratic members than in Paris, they attracted a mixture of adherents – unsuccessful doctors hoping to expand their clientele, radicals aiming to revamp the medical system, politicians seeking the support of disillusioned voters, social reformers with utopian visions of an egalitarian future, and mystics participating in international underground masonic networks.

Although linked by their involvement with animal magnetism, the activities of these provincial groups varied because they were run locally. Spiritualist cults were particularly strong in Lyons, where magnetisers were joined by Rosicrucians, Masons, Swedenborgians and other mystics and theosophists. These enthusiasts gradually dispensed with Mesmer's physical fluid, and sought to revitalise corrupted humanity by restoring a primitive spiritual realm. Far from dying out in the Revolution, such visionary mesmeric doctrines were developed in the 19th century by Charles Fourier, an influential social reformer who designed utopian communities where people would live together in harmony (although in practice, discord was rife in his authoritarian enclaves).

Mesmerism was given a different theoretical twist by Armand Puységur, who is the true hero of this story for many historians of psychoanalysis. A colonel from one of France's ancient families, Puységur had been trained by Mesmer at the Parisian Society of Harmony. Under his mesmerising influence, patients entered what might now be called an altered state of consciousness, and was known then as magnetic sleep or somnambulism. He claimed that he made his discovery accidentally, when a peasant suffering from congested lungs appeared to be asleep, but behaved as though he were awake. Realising that he responded to the mesmerist's suggestions, Puységur tried to send him pleasant thoughts: 'He then became calm – imagining himself shooting a prize, dancing at a party ... I made him move around a lot in his chair, as if dancing to a tune; while mentally singing it, I made him repeat it out loud ... The next day, no longer remembering my visit of the evening before, he told me how much better he felt.'[22] Although Mesmer stressed the physical bases of animal magnetism, Puységur insisted that healing the body also involved curing the mind.

Magnetic somnambulism sounds very similar to what is now called hypnosis, but its adherents regarded it differently. The Freudian model of the mind was undreamt of – even the word 'subconscious' had not yet been invented. Individual idiosyncrasies fascinate modern psychoanalysts, but to mystical mesmerists they seemed less significant than universal laws. When Puységur was invited to establish the Society of Harmony at Strasbourg, his host (who was a count) remarked to the distinguished colonel: 'Well, my dear Puységur, this is a real initiation, like our masonic induction ceremonies. Isn't animal magnetism a long-lost truth? And isn't the clarity achieved in the state of magnetic somnambulism the light to which we aspire?'[23]

* * * * *

Although Mesmer had retreated to a remote part of Germany, he never stopped boasting about his personal magnetic powers. Shortly before he died in 1815, Mesmer wrote yet another book outlining his ideas, describing how he stoked up his inner 'invisible fire' to magnetise other people. But by this time, practising mesmerists had abandoned their master's *baquet* and other magnetic equipment. They wanted to make mesmerism respectable by stripping away its association with radical politics and avoiding any hints of sexual impropriety.

First in France and then in Britain, magnetic somnambulism became important for controlling pain. Puységur had been much in demand for his ability to soothe away a toothache, but it was not until the 1830s that mesmerism was regularly used for operations. Anybody who needs convincing that surgery used to be agony should read Samuel Pepys describing how his troublesome bladder stone was extracted, or Fanny Burney's account of having her cancerous breasts removed. A Victorian campaigner for magnetic anaesthesia, a friend of Charles Dickens, was not exaggerating when he wrote: 'They who have heard, as I have heard, the dreadful shrieks, the sounds, more resembling the bellowing of a wild animal than the intonation of a human voice ... They, and they alone, can appreciate the almost overpowering thankfulness which swells the heart in return for a gift that, in a few moments, causes the shrill cry to sink into a tremulous murmur.'[24]

Mesmeric anaesthesia was eventually defeated by a later competitor: ether. This might seem like the straightforward victory of a superior technique over a quack gimmick – surely sympathetic surgeons would search out the most effective form of pain relief for their suffering patients? As often happens, actuality differed from ideology; ether was not – as one might perhaps expect – immediately adopted when it was first discovered, because the conflict between magnetism and ether was fuelled by the medical politics between rival groups of practitioners.

One problem was that inhaling gases seemed more appropriate for entertainment than for science. In Figure 28, Gillray is mocking Davy for his experiments with laughing gas, yet the scene reflects an actual episode when the flask had had to be torn away from an audience volunteer who was enjoying the experience too much. Gases were deemed to be for recreational purposes, not the medical operating theatre, and for 40 years countless patients survived surgery without them. When the anaesthesia wars were being waged in the 1840s, ether was being imbibed by self-indulgent medical students rather than being dispensed to needy patients.

In retrospect, the boundaries between respectable science and pseudo-science seem very clear. But for animal magnetism, there was no clear-cut way of drawing the lines between therapists who should be awarded the seal of approval and those who should be excluded. Franklin's report condemned Mesmer not because his treatments failed to work, but because he had no physical explanation for their success. But the ground rules kept changing, depending on who was laying them down. Half a century later, the advocates of ether could not account for its numbing effects, yet their evidence was accepted. In these asymmetrical trials, the testimony of the animal magnetisers was pronounced unsound because they were not fully qualified surgeons. Such scientific debates might appear to revolve around the search for truth, but they are also struggles for power.

FARADAY'S FIELDS

As the magnet selects from a quantity of matter the ferru-
ginous particles, which happen to be scattered through it,
without making an impression on other substances; so
imagination, by a similar sympathy, equally inexplicable,
draws out from the whole compass of nature such ideas as
we have occasion for, without attending to any others.

Alexander Gerard, *An Essay on Taste*, 1764

By 1812, Humphry Davy's rebellious youth was over.
Originally an apothecary's apprentice from Cornwall,
Davy was now a Sir and had married a wealthy widow. He
was also the butt of cruel jokes about his skill at social
climbing. Perhaps this explains why he acted swiftly and
sympathetically when he received a carefully bound copy
of handwritten notes on his own chemistry lectures at the
Royal Institution.

The lovingly prepared manuscript had been sent by
Michael Faraday, a blacksmith's son who was a trainee
bookbinder. Like Davy, Faraday was ambitious, and within
a few months Davy had secured him a job at the Royal
Institution. Faraday loved it. Although he was responsible
for all the menial tasks that nobody else wanted, he was
learning chemistry directly from Davy himself.

Within a few years, Faraday would become Europe's
leading expert on magnetism and electricity. By creating
ingenious devices to demonstrate that these two powers
of nature are ineluctably twined together, Faraday

founded a new science of electromagnetism and literally invented the motors of modern industry. Like Halley, he was fascinated by the patterns of terrestrial magnetism, and adapted Halley's terminology of lines and curves to develop a field theory of an electromagnetic influence extending throughout the whole of space.

Thanks to the collective efforts of Gowin Knight and countless other Enlightenment entrepreneurs, more openings were now available for men who had been born clever rather than rich; however, there was still no established career structure for men of science. Social relationships were what counted. Patronage was important, and Faraday depended on Davy's support. They were locked together in a traditional master–apprentice relationship, bound up with one another emotionally as well as scientifically. The situation was made more complicated by Davy's wife, who treated Faraday like a personal servant and kept reminding him about his working-class background. After Faraday had published some of his own research papers, he was elected a Fellow of the Royal Society. Faraday weathered Davy's fury, which was, perhaps, ignited by jealousy. The former assistant was becoming his teacher's peer, and Faraday is now far more famous than his patron.

The word 'scientist' was invented in 1833 to describe men like Faraday. This new term had become essential for bringing together under one umbrella researchers with very different interests – collectors classifying minerals and plants, physicists searching out the hidden powers of nature, chemists discovering more elements, mathematical astronomers predicting the location of undiscovered planets. Faraday became one of Britain's first professional research scientists, employed throughout his life by the Royal Institution, and in charge of waged assistants working with him in the Institution's laboratory.

Faraday was celebrated as a working-class hero, a classic case of rags-to-riches triumph. He took over Knight's magnetic machine (Figure 15, page 105) for his own experiments, but Knight also bequeathed a far more significant legacy – he helped to convince the state that science mattered. Faraday benefited from opportunities that had not existed when Knight, a gentleman's son with an Oxford degree, was struggling to make his own living from science. Knight had been forced to sell his scientific services to survive, but salaried Victorian scientists like Faraday could afford to think of themselves as being above such sordid matters. When the Prime Minister asked him if his electromagnetic machinery would ever be of any use, Faraday supposedly replied 'Why sir, there is every possibility that you will soon be able to tax it!'

Faraday's sardonic prediction proved right: we now depend on the giant electromagnetic industry he spawned, which drives our scientific consumer society. Yet that is not our only inheritance from the past. Mesmer was excommunicated from the science that Faraday was welcomed into, but thanks to him, magnetism's ancient sexualised vocabulary survives. A few years ago, a theatre reviewer summarised a play's plot as 'an intense ballet of attraction and repulsion' between a young lady and the family servant. 'They are', he told his newspaper readers, 'magnetised to each other in a lusting/loathing way until the famous offstage copulation makes the force-field between them go haywire.'[1] The force-field is Faraday's, but the metaphor is mesmeric.

NOTES

Mysterious Magnets

1 Samuel Pepys, *Diary* entries for 21 January 1665, 14 November 1666, 21 and 30 November 1667, 8 August 1666.

2 Stephen Pumfrey, *Latitude and the Magnetic Earth: The True Story of Queen Elizabeth's Most Distinguished Man of Science* (Cambridge: Icon Books, 2002).

3 D. Campbel [D. Defoe], *The Friendly Dæmon, or the Generous Apparition; Being a True Narrative of a Miraculous Cure, Newly Perform'd Upon that Famous Deaf and Dumb Person* (London, 1726).

4 J. Trenchard, *The Natural History of Superstition* (1709), pp. 26–30.

5 D. Dubois, *The Magnet: A Musical Entertainment* (London, 1771), p. 29.

6 T. and J. Holland, *Exercises for the Memory and Understanding, with a Series of Examinations* (Bolton, 1805), pp. 10–11.

Part One: Halley's Holistic Hypothesis

1 R.D. Altick, *The Shows of London* (Cambridge, MA: Harvard University Press, 1978), p. 72 (James Cox).

2 F. Sherwood Taylor, 'An Early Satirical Poem on the Royal Society', *Notes and Records of the Royal Society*, 5 (1947), pp. 37–46.

3 Thomas Hearne, quoted in A. Cook, *Edmond Halley: Charting the Heavens and the Seas* (Oxford: Clarendon Press, 1998), p. 102.

4 E. Halley, 'A Theory of the Variation of the Magnetical Compass', *Philosophical Transactions*, 13 (1683), pp. 208–21, at pp. 208–9.

5 John Locke, quoted in the 1785 edition of Samuel Johnson's *Dictionary*.

6 D. Defoe, *A General History of Discoveries and Improvements, in Useful Arts, Particularly in the Great Branches of Commerce, Navigation, and Plantation, in all Parts of the Known World* (London, 1727–8), pp. 298–9.

7 Letter of 1785 to Johann Sturm, reproduced (in Latin) in E.F. MacPike, *Correspondence and Papers of Edmond Halley* (Oxford: Clarendon Press, 1932), pp. 55–6. E. Halley, 'A Discourse concerning Gravity, and its Properties, wherein the Descent of Heavy Bodies, and the Motion of Projects is Briefly, but Fully Handled: Together with the Solution of a Problem of Great Use in Gunnery', *Philosophical Transactions*, 16 (1686), pp. 3–21, at p. 5.

8 E. Halley, 'Some Remarks on the Variations of the Magnetical Compass Published in the Memoirs of the Royal Academy of Sciences, with Regard to the General Chart of those Variations made by E Halley; as also concerning the True Longitude of the Magellan Streights', *Philosophical Transactions*, 29 (1714), pp. 165–8, at p. 165.

9 S. Schaffer, 'Halley's Atheism and the End of the World', *Notes and Records of the Royal Society*, 32 (1977), pp. 17–40.

10 Halley, 'Account of the variation of the magnetical needle', p. 575.

11 Halley, *ibid*, p. 575, p. 577.

12 Quoted in S. Snobelen, 'On Reading Isaac Newton's *Principia* in the 18th Century', *Endeavour*, 22 (1998), pp. 159–63, at p. 159.

13 W. Whiston, *The Longitude and Latitude found by the Inclinatory or Dipping Needle; Wherein the Laws of Magnetism are also Discover'd* (London: J Senex, 1721), especially pp. 44–75.

14 W. Stukeley, *Stonehenge a Temple Restor'd to the British Druids* (London: 1740), pp. 56–66.

15 J. Savery, 'Account of the Savery Family', British Library Additional Manuscripts 44058.

16 Royal Society Journal Book, vol. 13, pp. 456, 466, 494 (16 April to 11 June 1730).

17 S. Savery, 'Magnetical Observations and Experiments', *Philosophical Transactions*, 36 (1730), pp. 295–340, at pp. 333–40.

18 S. Savery, British Library Additional Manuscripts 4433, ff. 62–87, at f. 67r.

19 P. Cockburn, *An Enquiry into the Truth and Certainty of the Mosaic Deluge* (London: 1750), p. 311.

20 *The Weekly Journal*, 10 March 1716.

21 E. Halley, 'An Account of the Late Surprizing Appearance of the Lights seen in the Air, on the Sixth of March last; with an Attempt to Explain the Principal Phænomena thereof', *Philosophical Transactions*, 29 (1716), pp. 406–28, at p. 406, p. 415.

22 Halley, *ibid*, p. 419, p. 421.

23 Halley, *ibid*, p. 421.

24 Halley, *ibid*, p. 428.

25 Halley, 'Account of the variation of the magnetical needle', p. 575.

26 R. Bentley, *A confutation of atheism from the origin and frame of the world* (London, 1693), p. 6.

27 C. Mather, *The Christian philosopher: A collection of the best discoveries in nature, with religious improvements* (London, 1721), p. 19, pp. 109–10.

28 Halley, *ibid*, p. 576.

29 L. Holberg, *A Journey to the World Under-ground. By Nicholas Klimius* (London, 1742), p. 318.

30 John Barrow, quoted in Marilyn Butler's introduction to M. Shelley, *Frankenstein, or The Modern Prometheus: 1818 Text* (Oxford and New York: Oxford University Press, 1993).

31 James Gregory, quoted in Cook, *Halley*, p. 260.

32 MacPike, *Correspondence*, p. 243.

33 Cook, *Halley*, p. 269.

34 E. Harrison, *Idea Longitudinis: being, a Brief Definition of the Best Known Axioms for Finding the Longitude* (London, 1696), p. 33, p. 30.

35 Halley quoted in Armitage, *Halley*, p. 143.

36 Letters to the Admiralty reproduced in N.J. Thrower, *The Three Voyages of Edmond Halley in the Paramore 1698–1701* (London: Hakluyt Society, 1981), pp. 306–8.

37 E. Halley, 'Some Remarks on the Variations of the Magnetical Compass Published in the Memoirs of the Royal Academy of Sciences, with Regard to the General Chart of those Variations made by E Halley; as also concerning the True Longitude of the Magellan Streights', *Philosophical Transactions*, 29 (1714), pp. 165–8, at p. 166.

38 J. Fergusson, *Observations on the Present State of the Art of Navigation, with a Short Account of the Nature and Regulations of a*

Society now Forming for its Effectual Improvement (London, 1787), pp. 7–8.

39 Royal Society Journal Book, vol. 19, p. 366 (30 November 1747).

40 D. Defoe, *A General History of Discoveries and Improvements, in Useful Arts, Particularly in the Great Branches of Commerce, Navigation, and Plantation, in all Parts of the Known World* (London, 1727–8), p. 303.

Part Two: Knight's Navigational Novelties

1 P. Toynbee and L. Whibley, *Correspondence of Thomas Gray* (Oxford: Clarendon Press, 3 vols, 1935), vol. 2, p. 632.

2 Letter from Peter Collinson to Cadwallader Colden of 26 April 1745, reproduced in C. Colden, *The Letters and Papers of Cadwallader Colden* (New York: New York Historical Society, 9 vols, 1917–23), vol. 3, p. 114.

3 E. Gibbon, *The Autobiography and Correspondence of Edward Gibbon, the Historian* (London: Alex Murray, 1869), p. 24.

4 Letter from Cook to the Admiralty Board of 25 July 1768, PRO: ADM 1/1609.

5 Franklin, letter to Ezra Styles of 10 July 1755, reproduced in L.W. Labaree and W.B. Willcox, *The Papers of Benjamin Franklin* (New Haven: Yale University Press, 1960–83), vol. 6, p. 103. T.H. Croker, *Experimental Magnetism, or, the Truth of Mr Mason's Discoveries in that Branch of Natural Philosophy, that there can be no such Thing in Nature, as an Internal Central Loadstone, Proved and Ascertained* (London: J. Coote, 1761), p. 8.

6 B. Wilson, 'Autobiography', 1783, unpublished typescript, National Portrait Gallery, pp. 17–18.

7 W. Mountaine and J. Dodson, *An Account of the Methods Used to Describe Lines, on Dr Halley's Chart of the Terraqueous Globe; Shewing the Variation of the Magnetic Needle about the Year 1756, in All the Known Seas; Their Application and Use in Correcting the Longitude at Sea, with some occasional Observations relating thereto* (London: Mount and Page, 1758), p. 12.

8 Letter to Wilson of 24 July 1746, British Library Additional Manuscripts 30094, f. 22.

9 E. Stone, *The Construction and Principal Uses of Mathematical Instruments Translated from the French of M Bion* (London, 1758), p. 307.

10 Mary Montagu's letter to Alexander Pope of 10 October 1716, reproduced in A. Pope, *The Correspondence of Alexander Pope* (Oxford: Clarendon Press, 1956), vol. 1, pp. 365–6.

11 N. Ward, *The London-Spy Compleat* (London: Casanova Society, 1924), pp. 59–60.

12 Undated letter, British Library Additional Manuscripts 30094, f. 91.

13 J. Boswell, *Boswell's Life of Johnson* (London: Oxford University Press, 1927), vol. 1, p. 232.

14 J. Hervey, *Theron and Aspasio* (London, 1813), vol. 2, pp. 233–6.

15 *Universal Magazine* 6 (1747), p. 118.

16 W. Hutchinson, *A Treatise of Practical Seamanship* (1777), pp. 105–6.

17 J.C. Beaglehole, *The Journals of Captain James Cook* (Cambridge: Hakluyt Society, 1967–9), vol. 1, p. 138.

18 W. Burney, *A New Universal Dictionary of the Marine* (London, 1815), pp. 98–9.

19 M. Folkes, Royal Society Journal Book 19 (1747), p. 366.

20 W. Maitland, *An Essay towards the Improvement of Navigation, chiefly with respect to the Instruments Used at Sea* (London: for the author, 1750), p. 31.

21 J. Waddell, 'A letter concerning the effects of Lightning in destroying the Polarity of a Mariners Compass', *Philosophical Transactions*, 46 (1749), pp. 111–12.

22 W.C. Lukis, *The Family Memoirs of the Rev William Stukeley, MD and the Antiquaries and Correspondence of William Stukeley, Roger and Samuel Gale, etc* (Durham: Surtees Society, 1882–7), vol. 2, pp. 361–2.

23 W. Maitland, *An Essay towards the Improvement of Navigation, chiefly with respect to the Instruments Used at Sea* (London: for the author, 1750), p. 29.

24 W. Hutchinson, *A Treatise of Practical Seamanship* (1777), pp. 104.

25 Logbook of the *Fortune*, September 1751, PRO: ADM 51/361.

26 Letter from the Navy Board to the Admiralty Board of 29 April 1752, PRO: ADM 106/2186, f. 175.

27 Letter from Cook to the Navy Board of 8 March 1768, PRO: ADM 106/1163.

28 G. Robertson, *The Discovery of Tahiti: a Journal of the Second*

Voyage of HMS Dolphin round the World, under the Command of Captain Wallis RN, in the Years 1766, 1767, and 1768 (London: Hakluyt Society, 1948), p. 29.

29 John Pringle, 'A Discourse on the Different Types of Air', *Philosophical Transactions*, 63 (1774), p. 28 (appendix).

30 J. Purdy, *Memoir, Descriptive and Explanatory, to Accompany the New Chart of the Atlantic Ocean* (London: R.H. Laurie, 1829), p. 337.

31 J.T. Desaguliers, *A Course of Experimental Philosophy* (London, 1734), p. vi.

32 B. Langrish, *A New Essay on Muscular Motion* (London: for A. Bettesworth and C. Hitch, 1733), p. 81.

33 P. Templeman, *Curious Remarks and Observations in Physics, Anatomy, Chirurgery, Chemistry, Botany and Medicine. Extracted from the History and Memoirs of the Royal Academy of Sciences at Paris* (London: C. Davis, 1753), p. 173.

34 G. Knight, *An Attempt to Demonstrate, that all the Phœnomena in Nature may be Explained by Two Simple Active Principles, Attraction and Repulsion: Wherein the Attractions of Cohesion, Gravity, and Magnetism, are Shewn to be One and the Same; and the Phœnomena of the Latter are more particularly Explained* (London, 1748), p. 79.

35 I. Newton, *Opticks: or, a Treatise of the Reflexions, Refractions, Inflexions and Colours of Light* (Dover Publications: New York, 1952), p. 267, pp. 375–6 (Query 31).

36 J. Bate, *Experimental Philosophy Asserted and Defended, against some Late Attempts to Undermine it* (London: J. Bettenham, 1740), pp. 9–10.

37 Knight, *Attraction and Repulsion*, p. 47.

38 Knight, *Attraction and Repulsion*, pp. 82–3.

39 Arthur Murphy (1793) quoted in J.C.D. Clark, *Samuel Johnson: Literature, Religion and English Cultural Politics from the Restoration to Romanticism* (Cambridge: Cambridge University Press, 1994), p. 34.

40 J. Cook, *Clavis Naturæ: or, the Mystery of Philosophy Unvail'd* (London: for W. Meadows, 1733), p. 56.

41 G. Adams, *Lectures on Natural and Experimental Philosophy, Considered in it's Present State of Improvement* (London: R. Hindmarsh, 1794), p. 436.

Part Three: Mesmer's Magnetic Medicine

1 Quoted in R. Darnton, *Mesmerism and the End of the Enlightenment in France* (Cambridge, MA: Harvard University Press, 1968), p. 66.

2 T. Brown, *The Third Volume of the Works of Mr Thomas Brown, Serious and Comical in Prose and Verse* (London, 1720), p. 100.

3 T. Cavallo, *A Treatise on Magnetism, in Theory and Practice, with Original Experiments* (London: for the author, 1787), p. 103.

4 G. Bloch, *Mesmerism: A Translation of the Original Scientific and Medical Writings of F.A. Mesmer* (Los Altos: William Kaufmann, 1980), p. 38.

5 G. Bloch, *Mesmerism: A Translation of the Original Scientific and Medical Writings of F.A. Mesmer* (Los Altos: William Kaufmann, 1980), p. 89.

6 Quoted and translated (I have made slight alterations) in L. Wilson, *Women and Medicine in the French Enlightenment: The Debate over* Maladies des Femmes (Baltimore and London: Johns Hopkins University Press, 1993), pp. 104–5.

7 Quoted in V. Buranelli, *The Wizard from Vienna* (London: Peter Owen, 1975), pp. 110–11.

8 Letter of 5 December 1787 from Catherine Wright to William Withering (about John de Mainauduc), Royal Society of Medicine, London.

9 M.M. Tinterow, *Foundations of Hypnosis: From Mesmer to Freud* (Springfield: Charles C. Thomas, 1970), p. 98 (the entire report is reproduced).

10 Tinterow, *Foundations of Hypnosis*, p. 102.

11 Joseph Servan, quoted in Wilson, *Women and Medicine*, p. 115.

12 Tinterow, *Foundations of Hypnosis,* p. 88.

13 Quoted in S. Schaffer, 'Self Evidence', in *Questions of Evidence: Proof, Practice, and Persuasion across the Disciplines* (eds J. Chandler, A. Davidson and H. Harootnian; Chicago: University of Chicago Press, 1994), pp. 56–91, at p. 82.

14 S. Coleridge, *Lectures 1795 on Politics and Religion* (London: Routledge and Kegan Paul, 1971), p. 328.

15 Quoted in A. Crabtree, *From Mesmer to Freud: Magnetic Psychology and the Roots of Psychological Healing* (New Haven and London: Harvard University Press, 1993), p. 93.

16 Dr Pujol, quoted in Wilson, *Women and Medicine*, p. 120.

17 J. Robison, *Proofs of a Conspiracy against all the Religions and Governments of Europe, Carried on in Secret Meetings of Free Masons, Illuminati, and Reading Societies* (London: for T. Cadell and W. Davies, 1797), p. 6.

18 Quoted in J.B. Beer, *Coleridge the Visionary* (London: Chatto and Windus, 1959), p. 152.

19 Anon., *The Sceptic* (Retford: for West and Hughes, 1800), pp. 1–11.

20 Southey quoted on p. 48 of G. Myers, 'Fictionality, Demonstration, and a Forum for Popular Science: Jane Marcet's *Conversations on Chemistry*', in B.T. Gates and A.B. Shteir (eds), *Natural Eloquence: Women Reinscribe Science* (Wisconsin: University of Wisconsin Press, 1997), pp. 43–60.

21 Quoted in Darnton, *Mesmerism*, p. 91, p. 96.

22 Quoted in Crabtree, *From Mesmer to Freud*, p. 38.

23 Quoted in Crabtree, *From Mesmer to Freud*, p. 70.

24 Chauncey Townshend, quoted in A. Winter, *Mesmerized: Powers of Mind in Victorian Britain* (Chicago: University of Chicago Press, 1998), p. 170.

Faraday's Fields

 1 Paul Taylor, Independent, 22 March 1991, p. 17 (on August Strindberg's Miss Julie).

FURTHER READING

Fatal Attraction has been designed to complement three other Icon books on science and create a four-volume complete introduction to the history of electromagnetism. Because my account deals with magnetism in the Enlightenment, one tailor-made companion volume is my own *An Entertainment for Angels: Electricity in the Enlightenment* (Cambridge: Icon, 2002). To understand what happened before and after, there are also two other essential books to read. The first is Stephen Pumfrey's *Latitude and the Magnetic Earth: The True Story of Queen Elizabeth's Most Distinguished Man of Science* (Cambridge: Icon, 2002), which provides an excellent guide to William Gilbert and 17th-century magnetism. The other is Iwan Morus's *Michael Faraday and the Electrical Century* (Cambridge: Icon, 2004), an expert's easy-to-read account of how Michael Faraday joined electricity and magnetism together in the 19th century.

When I was writing *Fatal Attraction*, I relied on the information from many superb books. If you are interested in following up a particular topic, then I suggest you start with one of these:

Cook, Alan, *Edmond Halley: Charting the Heavens and the Seas* (Oxford: Clarendon Press, 1998). The fullest scientific biography.

Crabtree, Adam, *From Mesmer to Freud: Magnetic Sleep and the Roots of Psychological Healing* (New Haven and London: Yale University Press, 1993). Detailed but well-written study establishing Mesmer as the forefather of psychoanalysis.

Darnton, Robert, *Mesmerism and the End of the Enlightenment in France* (Cambridge, MA: Harvard University Press, 1968). Highly readable: now the classic revisionist version.

Fara, Patricia, *Sympathetic Attractions: Magnetic Practices, Beliefs, and Symbolism in Eighteenth-Century England* (Princeton: Princeton University Press, 1996). The only scholarly study placing Enlightenment magnetism in its full cultural context.

Heilbron, John L., *Elements of Early Modern Physics* (Berkeley, Los Angeles and London: University of California Press, 1982). A full academic study of electricity in the 18th century.

Home, Roderick W., 'Introduction', in Roderick Home and P.J. Connor (eds), *Æpinus's Essay on the Theory of Electricity and Magnetism* (Princeton: Princeton University Press, 1979), pp. 137–88. A detailed study of magnetic science in the 18th century.

Porter, Roy, *Enlightenment: Britain and the Creation of the Modern World* (London: Penguin, 2000). Unparalleled survey of the British Enlightenment.

Winter, Alison, *Mesmerized: Powers of Mind in Victorian Britain* (Chicago: University of Chicago Press, 1998). Wide-ranging, original and beautifully written study of mesmerism in the 19th century.

An Entertainment for Angels
Electricity in the Enlightenment
Patricia Fara

NOW OUT IN PAPERBACK!

'Vividly captures the ferment created by the new science of the Enlightenment … Fara deftly shows how new knowledge emerged from a rich mix of improved technology, medical quackery, Continental theorising, religious doubt and scientific rivalry.' *New Scientist*

'A concise, lively account.' Jenny Uglow, author of *The Lunar Men*

'Neat and stylish … Fara's account of Benjamin Franklin's circle of friends and colleagues brings them squabbling, eureka-ing to life.' *Guardian*

'Combines telling anecdote with wise commentary … presents us with numerous tasty and well-presented historical morsels.' *Times Higher Education Supplement*

Electricity was the scientific fashion of the Enlightenment. Patricia Fara tells the engrossing tale of the strange birth of electrical science – from a high-society party trick to a symbol of man's emerging dominance over nature.

From innovator-turned-statesman Benjamin Franklin to Romantic poet Shelley, the ability to harness electricity was a true revolution for human science – and, as one look around the technological world we in the West inhabit makes clear, for human society.

UK £6.99 • Canada $15.00 • USA $9.95
ISBN 1 84046 459 3

Latitude & The Magnetic Earth
The True Story of Queen Elizabeth's Most Distinguished Man of Science
Stephen Pumfrey

NOW OUT IN PAPERBACK!

'Stephen Pumfrey provides a chunky read with much more to it than at first meets the eye. He marshals his scientific and philosophical themes impressively while adding flesh to the hitherto enigmatic Gilbert.' *New Scientist*

William Gilbert was the most distinguished man of science in England during the reign of Queen Elizabeth I. The first person to use the terms electric attraction, electrical force and magnetic pole, he is considered to be the father of electrical studies.

Gilbert's world was that of Elizabeth's royal court, a hive of elite mariners and navigators. Through them, he came to hear of a new discovery made by a retired sailor turned compass-maker, the magnetic 'dip'. Using some of the first examples of experimental method ever recorded, Gilbert came to consider the Earth as one vast spherical magnet, with the accompanying ability to determine much more accurately a ship's latitude at sea.

Lively and accessible, *Latitude & the Magnetic Earth* – the first new exploration of Gilbert for forty years – brings the story up to date, leaving the reader with a vivid feel not only for the conflicts surrounding Gilbert's discoveries and his scientific legacy, but for the man himself.

UK £6.99 • Canada $15.00 • USA $9.95
ISBN 1 84046 486 0

Sex, Botany and Empire

The Story of Carl Linnaeus and Joseph Banks
Patricia Fara

NOW OUT IN PAPERBACK!

'Absorbing' *Observer*
'Engrossing' *Good Book Guide*
'Tremendously impressive' Jenny Uglow
'A rollicking read' *New Scientist*

When the imperial explorer James Cook returned from his first voyage to Australia, the scandal writers mercilessly satirised the amorous exploits of his botanist, Joseph Banks, whose trousers were reportedly stolen while he was inside the tent of Queen Oberea of Tahiti.

Enlightenment botany was fraught with sexual symbolism: Carl Linnaeus's controversial new system for classifying plants was based on their sexual characteristics, and the dangerously gendered language of flowers resonated with erotic allusions. In Sweden and Britain, both imperial powers, Linnaeus and Banks ruled over their own small scientific empires, promoting botanical exploration to justify exploiting territories, peoples and natural resources. Regarding native peoples with disdain, these two scientific emperors portrayed the Arctic North and the Pacific Ocean as uncorrupted Edens enjoying a naïve sexual freedom.

Patricia Fara reveals the existence, barely concealed under Banks's and Linnaeus's camouflage of noble Enlightenment, of the altogether more seedy drives to conquer, subdue and deflower in the name of the British Imperial state.

UK £6.99 • Canada $15.00
ISBN 1 84046 573 5

Other Icon science titles available in paperback

The Discovery of the Germ
Perhaps the greatest single advance in the history of
medical thought
ISBN 1 84046 502 6 £6.99

Perfect Copy
Unravelling the cloning debate
ISBN 1 84046 380 5 £7.99

Eureka!
How science came about thanks to the ancient Greeks
ISBN 1 84046 374 0 £6.99

Knowledge is Power
How magic, the government and an apocalyptic vision
inspired Francis Bacon to create modern science
ISBN 1 84046 473 9 £6.99

The Manhattan Project
Big science and the atom bomb
ISBN 1 84046 504 2 £6.99

Moving Heaven and Earth
Copernicus and the solar system
ISBN 1 84046 251 5 £5.99

Harvey's Heart
The discovery of blood circulation
ISBN 1 84046 248 5 £5.99

Turing and the Universal Machine
The making of the modern computer
ISBN 1 84046 250 7 £5.99

Dawkins vs Gould
How the battle of evolutionary theory is still fought over,
some 130 years after Darwin first announced his theory
ISBN 1 84046 249 3 £5.99

Available in hardback

Watt's Perfect Engine
Steam and the age of invention revealing the warts-and-all
James Watt
ISBN 1 84046 361 9 £9.99

Constant Touch: A Global History of the Mobile Phone
A fascinating exploration of the mobile phone
ISBN 1 84046 419 4 £9.99

Lovelock and Gaia
How Gaia theory came about and its long struggle to gain
acceptance
ISBN 1 84046 458 5 £9.99

Kuhn vs Popper
The struggle for the soul of science
ISBN 1 84046 468 2 £9.99

The Autobiography of Charles Darwin
ISBN 1 84046 503 4 £9.99